The
Family Cow

Dirk van Loon

GARDEN WAY PUBLISHING
CHARLOTTE, VERMONT 05445

ILLUSTRATIONS BY ANNA DIBBLE

DESIGNED BY DAVID ROBINSON

Printed in the United States by Vermont Printing Co.

Library of Congress Cataloging in Publication Data

van Loon, Dirk.
  The family cow.
  Bibliography: p.
  Includes index.
  1. Dairying. 2. Cows. 3. Dairy cattle. I. Title.
SF239.V26   636.2'1'4   75–26148
ISBN 0–88266–066–7 pbk.

Contents

Introduction

A few months ago I asked an agricultural extension agent in New England about family cows in his county. The man snorted, laughed and pulled out a wide roll of pale green computer print-outs for farm censuses taken over the past twenty years.

"Family cow's a thing of the past," he said, and there was a note of triumph in his voice. He swept a broad, pink palm across page after curled page, stabbing dates and figures. "Look here . . . and here. There's your last family cow. 1958."

Innocently I asked why they had disappeared. The agent smiled. "There's no money in a family cow . . . . Only work." He warmed up to his theme, looked me sharp in the eye and challenged. "Fifteen tons of manure a year! You going to tell people that?"

I had thought of it, but not in the way he implied—gesturing with arms outstretched over a pile that in his own mind must have filled the office.

As for making money with a family cow, that never had been our object in buying one. We didn't want to lose money with her—and we didn't. On a cash basis, unless the cow suddenly dies, most people stand to come out a little ahead.

But on leaving that county agent's office that day I felt a little behind. It wasn't the "no money" or "all work" arguments. I'd heard them before. It was the "fact" that the family cow was extinct in this fairly typical rural corner of the United States.

There are hundreds of family cows in the Maritime Provinces. But suddenly I was confronted with the possibility that this might be a freak result of isolation and economic factors that didn't hold with the rest of North America. If the family cow was gone, it had to be because there was something very impractical about her. I didn't know what it was. But how could I write a practical book on an impractical subject?

I had to find a family cow! Not having lived in the area for many years, I didn't know where to start looking.

A few minutes later I was in the next town having coffee with an old friend. She's not a farmer, but I told her my problem. "Oh," she replied. "How about the Bandleaders? They've got one. And then there's the Appleblossoms." I drove to the next town, stopping by the meat market to pick up some hamburger. Did they know of any family cows? They knew of four more.

What an ignorant computer! But no, it was my ignorance in forgetting that computers can only count what a technician feeds them. *Technician does not think family cows are practical? Technician engineers instant extinction.* So the family or "house" cow, as she is also called, is not gone. She is only sometimes forgotten.

I mention the incident because it is not unusual to find people in professional agriculture who are down on the idea of the family cow. Why I don't know, because often the same person will be happy to offer assistance without argument or judgment to anyone wanting to raise vegetables or fruits or bees, goats, pigs, horses or even beef cows.

But don't be discouraged. Knock on the next door. I did, and over the past couple of years I've found many other professional agriculturalists who were willing to jump out of their way to help me write this book.

Seventy-five years ago I don't think a book of this kind would have been needed. For one thing single- and two-cow dairies were as common as dirt roads throughout rural Canada and the United States. It would have been hard to find neighbors who didn't have a wealth of first-hand knowledge about cows.

Then too, the full force of twentieth century science and technology didn't hit North American agriculture until after World War I. Until that happened much the same skills went into keeping one cow as went into keeping a commercial herd. One book of dairy farming would have suited everyone's needs.

Books about dairy cows written in the past 20 years or so reflect huge changes that have come to commercial milk farming since the two world wars. Many of the changes were direct results of the wars that stole labor forces and replaced them with machines; that pushed scientific discoveries, sometimes ahead of the intelligence to use them wisely.

Cows are easy to raise, but I don't think modern books on dairy cow management show it. They are geared for the commercial farmer struggling at the bottom of an economic heap dominated by industry and urban mentality, power and greed. They are highly technical books. But worst of all they don't begin at the beginning.

Most have been written by a person who assumed the reader was raised on a milk farm. It's an easy assumption to fall into when you figure that the $2,500-per-cow cost of going into commercial milk farming establishes inheritance as about the only way to get started.

It doesn't have to cost nearly as much to get started with a family cow. The cow can be cheaper because she doesn't have to be the big producer. Her barn doesn't have to be half as fancy as the barn required by law or economics on a commercial farm. It won't take the enormous expenditures in machines and fossil fuels to keep the family cow that it takes to keep cows on modern push-button farms.

It doesn't have to cost as much to feed the family cow because there is no need to push her toward the limit of her abilities. Cows are closely related to deer. They evolved as brush- and grass-eaters, and on good forages alone they can produce up to six or ten quarts of milk a day depending on the individual and the breed. Add a moderate supplement of grains and protein and they can produce that much milk or more every day for close to a year.

It will take more human effort and time to keep the family cow than it takes, per cow, to care for 60 or 100. It's on this basis that a lot of people decide that the family cow is not "economically feasible." That's fine if a person wants to direct his or her life on that kind of time-is-money philosophy. But that person had better board up the kitchen and the laundry. And by all means forget about vegetable gardening and canning and freezing, because these are all areas where the family is throwing money away.

The Family Cow has come from personal experiences with Gladys and friends and from the experiences of many people who have kept family cows all their lives. Dozens of agricultural extension services in the United States and Canada have helped with information intended to keep the book from being narrowly confined to what works in New England and the Maritimes.

The Family Cow is a book for beginners—people who may be several generations "off the farm." But it goes on from there, and sometimes a bit deeper, into the whys and hows of keeping a cow and calf through an average year. It is a gathering of old and new information. Some of the tables and graphs reproduced here are 30 years old or more, in which case the actual quantities shown may not be the same as would be found today if the experiments on which they are based were conducted again. All the same, they are of more than historical interest because the relationships demonstrated between cows, crops or cropping practices still hold.

Some discrepancies in measurements may come up for Canadian readers who are used to Imperial measures or who are now going metric. Wherever it is not stated measurements are ones most commonly used in the United States.

This book may take some of the surprises out of keeping a milk cow. It won't take them all. There will be things to learn that couldn't be included in a library on the subject. That's what's fun about raising animals. It takes your own bank of personal experiences. There is no end to adventure and learning when three living things come together— like you, your cow and a piece of land.

Good luck.

Acknowledgments

To all of you who helped me with *The Family Cow*—more thanks than words can say. I don't know where I would have got without the unqualified support of people like Dr. James G. Welch of the Department of Animal Sciences at the University of Vermont, Craig Wheaton-Smith of Dorset, Vermont, and Stu Gibson, Vermont's Extension Dairyman, who gave so much, so cheerfully that they inevitably were the ones I came to lean most heavily upon as the book progressed.

I feel particularly fortunate in having met Mr. Wheaton-Smith whose extensive library and personal background in bovine genetics, English dairying and U.S. beef production gave me opportunities to explore beyond the limits I had originally seen for the book.

Drs. Eugene and Jean Ceglowski, veterinarians from Rupert, Vermont, put a lot of time and thought into what would eventually become chapters on health, breeding and calf rearing. Thanks to Dr. L. M. Cock, Prof. W. C. Mathewson and other members of the staff at the Nova Scotia Agricultural College for getting me started last year.

And thanks to the many more I haven't named, family, friends and acquaintances who contributed pieces of the puzzle or who, through their patience and generosity, gave me the space and places to work.

Dirk van Loon

To Howard and Dot Tingley—
with thanks from me and Gladys

CHAPTER 1

How Now,
(One Brown Cow)?

You buy a cow and you have to get her home. If you haven't got a truck, try walking. That's what we did, Gladys and I, a long time ago. Actually we didn't walk all those seven miles in from the main highway. Gladys dragged me at a run the first mile and I dragged her at a crawl the last two.

We had stopped for a rest there where the road widens at what they call Bill's Turn when a car pulled over. The window rolled down and a friendly face smiled, saying nothing for a minute, then: "What in hell are you going to do with that cow?"

"Milk her," I replied, as if that answered the question, when in fact it only answered what I knew for a certainty. I'd been around other peoples' cows most of my life. But being their cows, they had been making the decisions about what was to be done and when. There's a big difference between following directions and setting your own. What was I going to do with a cow? When? Why? How now, brown cow?

First of all there are general things to know about cows. Some I already knew. For instance, how it is that among domestic animals in North America cows, goats, sheep, geese and rabbits are the only ones that can take a diet that is mostly grass and turn it into food for humans. Being grazers (grass) and/or browsers (brush) they can "work" a piece of rocky, hilly, dry or wet ground that isn't suited to cultivation. I had plenty of that.

1

Goats versus Cows

If it is milk that is wanted, the choice here and today is between cows and goats. (There has been some recent bleating about milking sheep. Interesting. But goats and cows have been and probably will continue for some time to be the milk animals of North America.)

Between the two the choice will depend mostly on personal pref-

AVERAGE COMPOSITION OF MILK
OF DIFFERENT MAMMALS

SPECIES	Fat (percent)	Protein (percent)	Lactose (percent)	Ash (percent)	Total Solids (percent)
Antelope (Pronghorn)	1.3	6.9	4.0	1.30	25.2
Ass	1.2	1.7	6.9	0.45	10.2
Bison	1.7	4.8	5.7	0.96	13.2
Bitch	8.3	7.5	3.7	1.20	20.7
Buffalo	7.6	3.8	4.9	0.78	17.0
Camel	4.9	3.7	5.1	0.70	14.4
Cat	10.9	11.1	3.4	—	—
Cow	4.5	3.8	4.9	0.72	13.9
Dolphin	14.1	10.4	5.9	—	—
Elephant	15.1	4.9	3.4	0.76	26.9
Ewe	5.3	5.5	4.6	0.90	16.3
Goat	3.5	3.1	4.6	0.79	12.0
Guinea Pig	3.9	8.1	3.0	0.82	15.8
Kangaroo	2.1	6.2	—	1.20	9.5
Mare	1.6	2.7	6.1	0.51	11.0
Monkey	3.9	2.1	5.9	2.60	14.5
Rabbit	12.1	11.4	1.8	—	—
Rat	14.8	11.3	2.9	1.50	31.7
Reindeer	22.5	10.3	2.5	1.40	36.7
Seal	53.2	11.2	2.6	0.70	67.7
Sow	8.2	5.8	4.8	0.63	19.9
Vixen	6.3	6.3	4.6	0.96	18.2
Whale	34.8	13.6	1.8	1.60	51.2
Woman	4.5	1.1	6.8	0.20	12.6

erence, although there are distinct differences. Some people find goats' milk easier to digest, in part because the fat droplets are smaller. Because of these smaller droplets it is harder to separate the cream from goats' milk. Goats produce far less milk than cows. But then they eat far less and are less choosy about what they eat. Once I took care of a goat for a friend and thinking I would be doing the goat a favor I staked her in the middle of a patch of lush clover. Heck no. She strained and hauled until she turned the stake out of the ground to reach a scraggly thorn bush.

Goats are harder to fence, but some people never do fence a family goat (or cow), preferring instead to keep them on a stake and tether moved here and there as the feed is eaten down.

I think billy goats stink, and there's the end to that discussion as far as I should take it. There is a book on goats listed under *Other Books and Places* in the appendix.

The Work Involved

An average cow will produce 5,000 to 14,000 pounds of milk over 300 or so days following the birth of each calf. A cow lacking "persistency" might give half or less of these amounts because her system stopped producing milk after only a couple of hundred days. It's this way with beef cattle and probably was with all cows' wild ancestors.

But the dairy cow should continue at least 300 days if she is fed adequately and milked daily. Many go 400 days or more, and there is a record of a Holstein that gave milk for six years on a single calving.

It will take half to three-quarters of an hour a day to feed, water, milk and clean up after a cow, and perhaps a total of three or four hours a week to process the milk into skimmed milk, cream, butter and cottage cheese.

Making hard cheeses will take some more time, if these are wanted. There also will be time spent fixing fences or making repairs around the barn. If people make their own hay there will be a couple of weeks in early summer when that is the most important chore, though still not full-time. Come a thunder shower there's not a thing to be done but sit on your backside and cuss it out.

Ideal as raising a cow may sound, especially when the grapevine buzzes with the news there's a fine animal down the road for sale, cheap, there are some important questions to ask.

Should You Keep a Cow?

First, do you like cows? Is there something about the way they look or act that makes you feel you'd like to have one around?

Or is there something about the way they look or act—or smell—that says, "No, I really never liked cows very much."

These are instincts or deep-seated conditionings, and they can't be ignored. For success with cows a person has to like them, and not mind the smells, because that, too, will become a part of life, the degree depending on circumstances and the care taken with changes of clothes and so forth, to keep the smells in the barn.

Will local ordinances allow keeping a cow? The back side of main street may be knee-deep in horses in a village that is full of by-laws against cows, pigs, chickens, goats or what-have-you. These by-laws were written back when horses were cars. Since then horses have been the pleasure of those who run the town—or their children—and proposed laws against them have had a way of dying in committee.

I don't believe anyone with no domestic animals and the responsibilities that go with them should jump into owning a cow. It's not that they take so much time, but that in most cases they are an everyday, two-or-three-times-a-day obligation, especially when they are in milk.

Also, the cow makes the most sense when she is able to play at the hub of a coordinated whole that includes a garden—whose scraps she eats—and chickens and pigs that are able to make the most out of excess milk or milk by-products the family can't use.

I had thought of calling this book *The Neighborhood Cow,* for the reason that we have become a society of people who hate to be tied down. And unless there is at least one other person to step in on occasion to milk, feed and clean up after the cow, the idea of a day or weekend off can become an obsession.

There are ways to ease the burden, especially toward the latter part of the milking or lactation period. I'll mention them later. But for the most part it's a mistake for people to get a cow if they aren't settled down and willing to schedule their lives around the cow's needs.

Is there enough money on hand to risk at least the initial cost of buying a cow? A person could begin with an inexpensive calf, in which case there's little to worry about as far as losing a bundle all at once. But I think the best way to begin would be with a moderately-priced cow that is well along in a milking period and known to be with calf. She would cost about $200 or more, and of course there is no guarantee she won't get struck by a lightning bolt the next week.

Does It Pay?

About ten years ago the U.S. Department of Agriculture devised a simple accounting aimed at helping to decide whether it would "pay" to keep a family cow. Updated to today's prices it looks like this:

It will pay you to keep a cow if the cost of milk and butter your family needs is more than $1.30 per day:

If you buy a cow for	$300
And sell her in five years for	100
Cost for five years	$200

Cost for one year	$ 40
Interest on $300	30
Breeding charge	15
Grain mix, 1.5 tons	190
Three tons hay	240
	$515
Calf sells for	−20
Keeping cow one year costs	$495

Or $1.30 a day or about
the price of three U.S. quarts
of homogenized milk.

Granted this scheme doesn't take into account any medical bills or interest on capital costs or costs of barn and equipment. But neither does it consider the value of manure (about $30), or of quarts of milk sold, of which there should be many if she's any kind of a cow at all.

The sale of just one quart of milk a day at 40 cents a quart should cover the cost of grain for the year. A Vermont family turns its extra milk into yogurt that sells for 90 cents a quart.

The cow isn't necessarily going to go down in value. We bought Gladys for $175 and sold her three years later for $300.

For housing, a 10- by 15-foot shed with a tight roof and floor would do for a start in any climate, using canvas tarpaulins to cap piles of loose or baled hay.

Hay can be stored outdoors if need be. It's best to cover at least the peak of the stack with a canvas tarpaulin. Plastic is sometimes used, but even the heaviest plastic will begin to crack and tear after a few winter storms.

The amount of land needed for one cow and calf varies from a couple of hundred square feet to perhaps as many as 25 acres, depending on soil, climate and how a person plans to go about raising the animals.

On the average somewhere between five and ten acres of ordinary cleared land growing mixed grasses should be enough to provide pasture and winter roughages for the cow and calf in most areas of the United States and Canada.

Cows do not need much exercise. And so, although the words pasture and cow come to mind like a double exposure, the truth is that tons of milk are produced each year by cows that never step off a concrete lot. The system is called "zero pasturing" or "zero grazing" and involves trucking every bit of feed to the waiting cows.

The other extreme of land use would be to pasture a cow on the dry, sandy soils of the United States Southwest. There, the Colorado Agricultural Extension Service suggests 20 to 25 acres per cow on "ordinary" pasture. However, even here the figure could be halved or better through irrigation and management.

Some might say that the five- to ten-acre figure is crazy; that a cow and calf could be raised on half that. It could be done in many areas.

But in most cases it would take a few years of experience and good management because often whatever land is available has been used and abused for years.

Most family cow owners I have known farmed about eight acres. One family that had had a cow on the place all their lives got by beautifully with four. How they did it is described in Chapter 18 on land use.

You have to have shelter, land and permission from the mayor. What else? A wheelbarrow, a pitchfork, a manure fork, a shovel, hoe and rope. There is always need for another length of quarter- or half-inch rope around a barn. And a couple of sturdy pails. That's about it to start. It is better to feel your way into more and fancier gear, although again there will be more specific suggestions in later chapters.

CHAPTER 2

A History of Cattle

The domestication of cows began about 8000 years ago among people who had been living and evolving with wild cattle for ages. Some say it happened first in Western Asia. It wasn't long before there were many people there, in the Middle East and northern Africa who were keeping some cattle under some kind of control.

The taming of the wild ox, or *aurochs* as they came to be known in Europe, came after dogs, sheep and goats. The domestication of horses came later.

There wasn't a sign of a milk cow among earliest domestic cattle. In fact it was years before cows rated above goats, sheep, reindeer, water buffalo, camels or even horses as sources of milk or cheese, and then only in a few corners of the world.

Depending on whose system of classification is being followed, anywhere from two to four or five species or subspecies of the genus *Bos* gave way to domestication over the years.

One of the neatest theories holds that the two varieties of domestic cattle found in North America—those that are most common and the humped Brahmans or Zebus—are all one genus and species, *Bos taurus* (the Latin and Greek for ox), and that they came out of one wild ancestor called *Bos primigenius* (first ox). Supporting this theory is the fact that common and humped cattle breed back and forth without difficulty.

But other classifiers would give *taurus* the less specific status of subgenus, and say the two should be *Bos taurus typicus* (common or "European" cattle) and *Bos taurus indicus* (the "Indian" humped cattle). They point to the obvious differences between the two, particularly to the Brahman's fat and muscle hump, but also to its floppy ears and loose folds of skin and general form. They could ask, too, why it is that Brahmans are more heat-tolerant and why they are resistant to some tropical diseases that clobber other types of cattle.

There was a time when it was thought the Brahman's hump and skin folds had to be important for heat regulation, a thing developed over centuries through natural selection within a species of animal grown apart from all others that might once have claimed kinship. It is an idea the two-species theorists would like to believe.

But studies in recent years haven't been able to prove humps or skin folds do much for heat radiation. Into the breach jump the taxonomic "lumpers," suggesting now that all these differences could be nothing more than results of people-interference: people selecting for impressively high-shouldered animals; people wanting and selecting humps because they provided ideal support for neck yokes.

The end of this and many more arguments on the origin and proper division of cattle into special groups won't be heard for years. In many cases, especially when digging into the past, there is too little evidence to go on. Go far enough back and even the cow gets dinky. The earliest record of cow-like remains date back about 50 million years. She was the size of a fox terrier.

Coming back home, Brahman characteristics are rarely seen on North American farms outside of the Southern United States. Then they are more often seen in crosses with common beef breeds resulting in animals like the Brangus (Brahman times Angus) or the Santa Gertrudis (Brahman times Shorthorn). Brahmans could be used for milk. Some cows have had production records in the 8000 pounds-per-year range, which certainly would do for a family cow.

However, most family cows will be what I have been calling "common" types and they will be *Bos taurus* without question. Also we can look back to *Bos primigenius* as being most to blame for what we have.

Bos primigenius was usually black with perhaps a white patch on its forehead and a light stripe down shoulders and back. But there were variations from early on. Cave paintings of aurochs in Southern Europe that may be 25,000 to 30,000 years old have them black, brown, and sometimes spotted or roan.

It's thought cows were first domesticated to provide meat, hides and horns, either directly or by way of sacrifices, to humor the spirits

The aurochs, Bos primigenius. *Where it all began.*

who watched over wild oxen. This was terrific progress for people who, up to then, had had to rely on cave paintings and incantations to conjure up times of plenty—the way stockbrokers hope up-turned arrows and bold predictions will bring on their "bull" markets.

Certainly Stone Age people needed all the spiritual help they could get when they went hunting the aurochs with clubs and stone-tipped spears and arrows. These cattle towered over their gnat-like hunters and must have outweighed them by 1000 pounds or more. Often the bulls stood six feet at the shoulder behind horns spreading ten feet from tip to tip.

An outstanding trait of the aurochs, and one that may have led the way to the barn, was its habit of standing up to any threat. People admired the trait. They capitalized on it as well, since a standing target was always the most vulnerable to attack. The result over centuries may have been a mindless selection against the most aggressive aurochs bulls and cows, until eventually people whose ancestors followed migrating herds of wild cattle found themselves in the lead.

The last *Bos primigenius,* a cow, died in Poland in 1627. The spirit of these animals that had once roamed across the plains and forests of Asia, Africa and Europe, had been captured long before. The cow had become fixed in religion and mythology. The bull had become a standard for measuring human bravery and masculinity. Way before the time of Christ the great spectator sport on the island of Crete was watching young men and women pit acrobatic skills against the thrusting horns of wild bulls. Down through history the games have changed but the same human ideals are tickled by bull fights and bull-riding rodeo cowboys.

The earliest kept-cattle may have been runts. Some bone and fossil remains say that they were. Perhaps they became runted and stunted through starvation or conscious selection for smaller, more manageable animals.

**WHERE THE COW STANDS
IN AN ABBREVIATED CLASSIFICATION OF ANIMALS**

KINGDOM—*Animal*

PHYLUM—*Chordata*   Having a tubular nervous system

CLASS—*Mammalia*   Having hair and mammary glands

ORDER—*Primates*   (Humans and Monkeys)

ORDER—*Perissodactyla*   Odd-toed (horses, rhinoceroses)

ORDER—*Artiodactyla*   Even-toed

SUB-ORDER—*Suiformes*   Few upper incisors. (Pigs, having one-chambered stomachs)

SUB-ORDER—*Tylopoda*   Few upper incisors. (Camels, having three-chambered stomachs)

SUB-ORDER—*Ruminantia*   No upper incisors. Having so-called four-chambered stomachs

FAMILY—*Cervidae*   Most having bony antlers that are shed and renewed annually. (Deer, moose and caribou [reindeer])

FAMILY—*Antilocapridae*   Having branched, horny sheaths over bony cores. Only the sheath is shed each year (Pronghorn antelope)

FAMILY—*Bovidae*   Hollow horned, except in polled cattle where a lack of horns is caused by recessive genes for this trait. Horns, when present, are permanent (Cattle, sheep, goats, true antelopes, buffaloes, bison, yaks and musk oxen)

SUB-FAMILY—*Caprinae* Having high-crowned teeth

GENUS—*Ovis*   (Sheep)

GENUS—*Capra*   (Goats)

It's likely there wasn't much emphasis on the size or quality of domestic cattle back then. But from the beginning there may well have been pure status in having more cows than anyone else much as there is today among the pastoral Tutsi and Masai of Africa, among the gauchos of Argentina, and cowboys in our own West. Other Americans, North and South, would rather hang their status on money. But that only exposes a cultural taproot, because another word for money, "pecuniary," comes from *pecus,* the Latin for a herd of cows.

GENUS—*Ovibos*	(Musk oxen)
SUB-FAMILY—*Bovinae*	(Cattle, buffaloes, bison, yaks and others) On the basis of more subtle differences than went into earlier divisions, members of the *Bovinae* are further subdivided, but in varying ways depending on whose system of classification is being followed. One system would have them all members of:
GENUS—*Bos*	
	and would divide them further as:
SUB-GENUS—*Bubalus*	(Water buffalo)
SUB-GENUS—*Bibovus*	(Gaur, banteng)
SUB-GENUS—*Bisontus*	(Bison, yak)
SUB-GENUS—*Taurus*	(Cattle)
	with cattle being divided again into:
SPECIES—*typicus*	("European" cattle)
SPECIES—*indicus*	("Indian" or humped cattle)

The other system suggested in the text considers *Bubalus, Bibovus* and *Bisontus* as names designating other genera under the sub-family *Bovinae* and therefore on a par with *Bos. Taurus* then designates species, and *typicus* and *indicus,* if used at all, refer to sub-specific differences.

By 3000 B.C. domestic cattle had grown and evolved into sizes and types we know today, reflecting in part hugely improved abilities to feed captive animals. Artwork from this period shows cattle doing field work in Egypt and Mesopotamia. Milking had begun too, from the side in Egypt and from behind the cow in Mesopotamia. Cows are still milked from behind in some parts of the world.

Also by 3000 B.C. domestic cattle culture had gathered all the momentum it needed to push north into a new world that had been opening up as the last ice age receded. Nature is supposed to abhor a vacuum, and there was a heller that sucked whole plant and animal communities including wild cattle to the melting lips of the Wurm glacier that for some 25,000 years had covered thousands of square miles of northern Europe and Asia.

Farmers followed, driving herds of domestic cattle up river valleys and through mountain passes into eastern and central Europe. Other travelers drove their cattle aboard boats and sailed out the Mediterranean and up the coast to settle the British Isles, the Low Countries and on to Scandinavia. By 2000 B.C. there were probably few corners of Eurasia and Africa that didn't know at least the occasional tip, tap or crunch of a domestic cloven hoof.

Looking at today's domestic cattle, especially in Europe and North America where specialization has been taken to an extreme, it can be hard to picture the part cows played in these early movements of people from one part of the world to another.

Today there are special sacred cows and specially selected bulls for fights, rodeos and other spectaculars. There are more than 200 so-called Pure Breeds of meat, dairy, work or multipurpose cattle. Exactly how many is difficult to say from one year to the next, since they come and go with the human organizations behind them. But in the numbers there are breeds whose *average* cows produce 10,000 or more pounds of milk in less than a year, animals that can pull many hundreds of pounds, and beef breeds whose steers can mature and fatten in less than two years, yielding carcasses that will dress out better than 50 percent in marketable food. It's all quite wonderful.

But for most of their history the beauty of domestic cattle has not been that they were good for work, meat or to be cherished or milked. It has been in how well they suited whatever need was uppermost at the moment.

It was a universal cow that made it possible for people 5000 years ago and since, especially in temperate climates, to leap across miles of unknown territory in search of new lands. They provided fresh food for the journey. As pack animals they no doubt carried more than their own bodies over many a mountain pass. They provided food through the hard times while settlers were learning what their new land had to offer in the way of local game, fish and greens.

Colonial settlers in North America might harness the family cow to a burnt-land harrow. At the end of a day's scratching the cow could be parked in the barn and a kettle of milk drawn off for the supper table. It wasn't a terrible thing to do or unrealistic. Experiments in England and India have shown that working a well-fed milk cow four or five hours a day won't cause an appreciable drop in milk production.

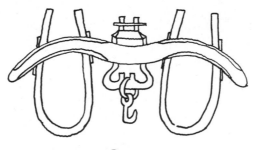

Oxen

For the long, hard grind there was and is the ox, a mature, castrated bull. In her *Handbook of Nature Study* Anna Botsford Comstock wrote, "For logging, especially in pioneer days, oxen were far more valuable than horses. They are patient and will pull a few inches at a time if necessary, a tedious work the nervous horse refuses to endure."

Another value over the horse she didn't mention was brought home to me a couple of years ago in Nova Scotia. An elderly farmer was anxious to show off the ten-year-old ox he'd raised from a calf. We went into the barn where the old man stepped up behind the animal that had so faithfully plowed his fields and hauled his firewood those many years. He gave the big ox an affectionate pat on the rump and smiled. I thought it was pride, and well-deserved too, for the years of work hardly showed in the animal's huge, muscled frame and sleek coat. But no. The farmer's eyes twinkled as he gave the ox another pat. "There's some fine steaks, you."

Storing loose hay at the Ross Farm Museum. A summer afternoon at an operating farm museum like this one at New Ross, Nova Scotia, is worth a library of books on nineteenth-century farming. Many of the practices demonstrated will be more applicable to part-time farming than those on a modern milk farm. Few part-time farmers will use oxen for farm power but the family cow could be trained for light draft work.

While universal cattle were helping to spread people across the globe, they carried one other basic quality with them—refinability—ready to show itself whenever the need or opportunity arrived.

Some of the history of refinement can be traced through language. Early in the history of domestication the word for cattle referred to all goods, merchandise or stock. Later it came to mean livestock. Then cows and horses were called large cattle and things like pigs, sheep and goats were small cattle. *Neat,* as in neat's foot oil, was the word later used in England for large cattle of a bovine ilk. It's said to have come from *naut,* the Icelandic word for ox. Later on in England milk or "milch" cows were known as "kine."

In the earliest years of domestication there was emphasis toward meat animals. Then work took over in many cultures around 5000 years ago when the yoke was invented. As pack or draft animals, cattle were *the* beasts of heavy burden for many years. Horses were only good for riding, packing and light draft until the neck collar was invented in China sometime in the first century before Christ. It took almost 1000 years for the invention to reach Europe.

Milk Cows

Although cows were being milked by about 3000 B.C., there is no way of knowing how good these cows were. I don't know of a milk record of any kind before about 350 B.C. when Aristotle found cows in northwestern Greece that were giving 32 quarts (64 pounds) a day. How many days in a row that kind of production went on he didn't say.

Another blip on the dairy scene comes from Pliny in the first century A.D., reporting that 1000-pound wheels of cheese were rolling out of the Po valley. This is an area of year-round pastures, but even so there must have been some fancy dairy animals up there. They probably were forerunners of the Lombardy cattle imported by the Dutch about 1200 A.D. By 1500 the fame they'd given the Po valley had moved with them to the Low Countries.

The first cows in the New World were brought to the Canadian Maritimes by Vikings about 1004 A.D. There were large dairy farms on Greenland at that time. Archeologists have discovered the ruins of one barn that had stalls for more than 100 cows.

Thorfinn Karlsefni is said to have brought dairy "type" cows when he tried to settle in Nova Scotia—or perhaps it was the south shore of Labrador. Maybe if they had been more the universal cow, their Viking keepers would have made it in their new home.

As it was, the attempted settlement failed within a year, leaving it up to Columbus to bring in the first permanent bovine settlers. (There never were wild cattle in the Americas.) Columbus landed his tough and rangy cattle on the Island of Hispanola in 1493. It is possible that descendants of these and of other Spanish imports became the famous Texas longhorns.

French settlers in Nova Scotia and Quebec imported cattle from northern France in the 1620s. The English brought cows; the Dutch brought cows; the Germans, too. Everybody sent home for a cow.

Most Englishmen called a milker "Bossie . . . Come boss," no doubt a leftover from the old Latin, *bos*. Other English and Scottish settlers called her "Cusha, cusha . . . Cushy cow." Acadians transplanted to Louisiana called her "Cha, cha . . ." Among French Canadians there was, and is, "Cho, cho, cho?" which, I was told, works best when it's accompanied by the rattle of grain or potatoes in a bucket—kind of a universal call.

None of the cows or bulls brought to the colonies were registered purebred anything, since the idea of pure breeds in cattle wouldn't be

The Canadian cow. Weighs 1000 to 1200 pounds. Colored black, shading to brown, with lighter muzzle and udder. May have white on the udder, stomach and chest. No information is available on the average annual milk yield for this breed, but it is said to average 4.5 percent butterfat. The breed was developed in Quebec (society founded in 1895) out of descendants of early imports from northern France. The society stresses that these cows were developed out of animals representing years of natural selection in North America, whereas all other dairy breeds have come from cows and bulls imported from Europe during the past 100 or so years. The society also notes the meatiness of the Canadian cow. Therefore it might be considered dual-purpose rather than purely dairy in type.

The overall triangular or wedge shape of the dairy cow quickly distinguishes her from the blocky beef "type" (right). Dotted lines indicate how the dairy and beef types also differ in cross-section.

invented until the nineteenth century. Instead they represented dozens of directions the aurochs had gone in thousands of years of isolation, mixing and isolation, on islands or tucked away in mountain valleys.

Although all of them today would be classified triple-purpose (milk, meat and work animals), they no doubt showed some degree of individual character that reflected a country or region of origin. Perhaps there would be some coat or color difference, or perhaps a difference of size or conformation. Some might produce more milk and less cream than others.

But whatever divergence may have been going on in various corners of Europe came to a crashing halt here. The Eastern seaboard of North America and the Spanish Islands became an ultimate mix-master. Although attempts were made in some communities to control breeding by establishing a town bull, the greater reality was no control at all. Ships trading cattle up and down the coast were like hummingbirds in a patch of morning glories, pollinating back and forth until American cattle were distinguishable by their lack of distinction.

These "native" cattle were unremarkable by today's specialist standards. A cow might not produce more than 1000 to 3000 pounds of milk in a year. Bulls not wanted for breeding were castrated and trained to the yoke, or they were grass-fattened for slaughter at the age of three years or more.

They were durable cattle, though. And they suited the needs of people too busy clearing and cultivating wild land to devote the time and expense involved in large-scale cheese or butter manufacture. It was before pasteurization, refrigeration and rail transportation, those keys to the eventual boom in fluid milk and cream markets, and the attempts to develop single-purpose, high-production dairy cows to fill them.

Milk

Fluid milk was scarcely used outside of European farming districts until the early 1600s. Some people thought it wasn't safe to drink milk, and considering the conditions of the time and what little was known about contagious diseases these people were right.

Though ahead of milk, possibly because it was easier to store and transport, butter didn't become a widely popular food until the late 1200s. Prior to that time there had been city people who thought butter made a fine skin cream but was hardly something they would put on a dinner plate—not a screwy idea back then when milkmaids were famous for their unblemished faces in a world scarred by smallpox. The fact, later proved by Edward Jenner, was that many milkmaids were picking up a milder cowpox disease from their animals that made them immune to the more serious smallpox organisms.

So many discoveries, inventions and social changes of the eighteenth and nineteenth centuries lent themselves to the exploitation of milk that it is hard sometimes not to think a need for milk wasn't somehow behind it all. That Louis Pasteur's name is on the process that finally made milk safe adds to the illusion. Actually Pasteur never worked with milk. Wine fermentation was his springboard to fame, and it was only years later (the 1890s) that Theobald Smith and others took some of Pasteur's ideas and applied them to milk purification.

The Breeds

"Pure" breeding in cattle goes back to the 1760s when an English farmer named Robert Bakewell began applying ideas he had developed through breeding sheep to the perfection of beef cattle. His original stock came from among the hundred or more varieties of cattle then found around Great Britain and Ireland, all of which were roughly split into three groups depending on whether they were short-, middle- or long-horned.

Bakewell bred what he thought the best to the best, selecting with great care but without worrying whether sire and dam were closely related or not—father-daughter, daughter-son. It made no difference to Bakewell, though his neighbors might shake their heads, for this inbreeding was absolutely frowned on at the time.

These attitudes changed as it became obvious that Bakewell's cattle not only were better but were increasingly able to produce predictable and acceptable offspring. Bakewell's cattle and the breeding practices behind them became the forerunners of the Short Horn Pure Breed, loosely framed in the *George Coates Herd Book* of 1822. The breed was given final and assured protection from outside blood in

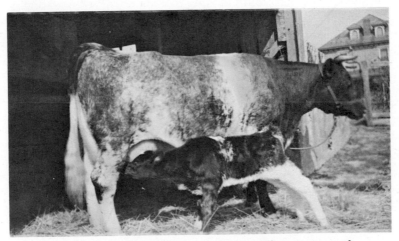

The Milking Shorthorn cow. Average weight, 1400 pounds. Colored red, red and white, white, or roan. The nose should be "flesh" colored, with no dark pigmentation. They yield 9,000 to 10,000 pounds of milk per year, with the milk averaging 4 percent butterfat. The breed comes from England and is descended from the oldest established breed, the Shorthorn (1822). The U. S. Milking Shorthorn Society was formed in 1920. Milking Shorthorns are the best known of the so-called dual purpose (milk and meat) breeds. Shorthorns and Milking Shorthorns may sometimes be called Durhams, a name that came with many of the early imports from the county Durham in northern England.

1876 when the Short Horn Society of Great Britain and Ireland incorporated the Coates book into their own, and from then on it was the official list of registered Short Horns.

The first of our dairy breeds began in 1868 with the publication of a herd book by the American Jersey Cattle Club, followed in 1872 by the formation of a club and publication of a herd book by breeders on the Island of Jersey.

Most of the breed associations like to promote the impression that their cows have been "pure" for many more years than is the case. For instance, the Jersey people say that in 1789 the Island of Jersey passed a law forbidding the importation of cattle. The implication is that, herd book or no, Jersey cows on the island haven't been violated by outsiders for close to 200 years. In fact, the law of 1789 only forbid the importation of cattle from France and was imposed for economic reasons.

The Holstein-Friesian Association (established 1871) talks of their breed being born in the Rhine Valley out of black cows and white cows brought to the region by the Bavarians and Fresians 2000 years ago. "The intermingling of these cattle evolved a black and white breed which, on the lush pastures of the region, developed size and producing-ability that gained them increasing fame and recognition as civilization advanced."

A pretty picture, but other paintings of cattle and descriptions of cattle grazing those lush pastures as recently as the early eighteenth century tell another story. The cattle were all sorts of colors, and mostly red and white. Some were even black and white. But this color pattern

AVERAGE U.S. YIELD OF MILK AND FAT

Breed	Number of Records	Milk (lbs.)	Fat (percent)	Fat (lbs.)
Ayrshire	22,336	11,610	3.87	450
Guernsey	93,392	10,285	4.58	471
Holstein	1,721,129	14,721	3.16	531
Jersey	103,053	9,497	4.94	469
Brown Swiss	27,111	12,743	3.98	508
Milking Short-horn	3,133	10,485	3.68	385

Based on official Dairy Herd Improvement Association test for calving year 1972, cows milked twice daily, 305-day lactations.

didn't become dominant until after an early eighteenth-century plague of rinderpest and pleuro-pneumonia just about destroyed the local cattle population of the area, forcing a large importation of cows and bulls from southern Denmark.

Early in the game of purebred dairy cows the pedigree was all important, and the longer the better. Since 1932, beginning with the Holstein-Friesian Association, the breeders have eased off on pedigree and placed much more emphasis on performance and progeny.

In spite of the advances in herd production averages, there may be miles to go before the genetic potential that lay buried in *Bos primigenius* is reached. Last year one registered Holstein in Pennsylvania produced 50,759 pounds of milk in 365 days, more than enough to have raised a whole herd of aurochs calves.

CHAPTER 3

Behaving Herself

"Boss? . . . Come bossie . . . Come boss."

It is late afternoon—milking time. The cows have been lying on the far side of the hill from the barn, chewing their cuds. Forty-thousand plus jaw movements a day for the average cow, according to one scientist studying rumination.

Eyesight

They are contented cows, with large brown eyes able to register the subtle colors and contrasts of reflected sunlight playing over the pasture and woods beyond. Cows see well in the dark. On some farms at times of the year when flies are thick, cows are pastured only at night and barned during the day so they will be freer from torment.

Cows and bulls are not enraged by the sight of any particular color. Certainly not by red, which, in fact, they may see as black since their visual spectrum is not the same as in humans. Bulls are just generally aggressive, especially if they are teased and provoked. The Spanish bulls of fight fame are chosen for their ferocity, and are provoked as well with spikes and lances. By the time the matador appears flapping his red cape the poor bull is ready to tackle anything. Only the flappiness helps draw the bull's attention from whatever other distractions might be in the ring. He is fooled into charging the cape, but not for long. They say after 10 or 15 minutes of stabbing cloth, bulls become "smart" and deadly dangerous, as they go after the fighters.

Bulls of the dairy breeds are, over all, more dangerous and un-predictable than beef bulls. It is in their blood and often aggravated by the fact that dairy bulls are kept unnaturally confined, while beef bulls may be allowed to run with a herd, perhaps even in the company of another bull or two which allows them a natural outlet for their aggressive interests.

Smell

The bovine sense of smell is keen. The smell of a cow in heat gets a bull riled. The cow doesn't have to be there, just the odors picked up on a person's clothing in passing through a nearby cowbarn will do it. Cows are supposed to hate the smell of blood. I have been around cows with pig blood on my boots and it didn't seem to bother them. Maybe it's only cow blood that upsets them.

Beads of moisture on the cow's smooth muzzle let her taste her food before drawing it into her mouth. This is helpful because cows are not able to spit. They can throw up, and sometimes heave violently like the business in babies that's called "projectile vomiting." Just as in babies, there is nothing to worry about if it only happens rarely.

Eating

In contrast to our own, a cow's tongue is rough and not as well equipped for tasting. It is long and powerful, and is used like a hand to reach out and grab hanks of grass or hay. In grazing, grass is drawn into the mouth, gripped between lower incisors and toothless upper front gums, and pulled from its anchoring stalks and roots.

Horses have incisors top and bottom which let them clip the grass almost at root level. They gather the grass with an almost prehensile upper lip—nip, nip. Sheep, too, use their lips for gathering and are able to graze closer to the ground than cows in spite of the fact that, like cows, they have no upper incisors. Also, because they have no upper front teeth, you needn't fear getting a horse-like nip from a cow or sheep. On the other hand a cow may try grazing a head of long hair.

Drinking

Cows, horses and sheep all drink alike, using their tongues as pistons to draw water up to where the muscles of swallowing can take over.

Cows also use their rough tongues for grooming. Being less agile, they aren't as good at it as dogs or cats, but their short hair tends to shed dirt anyway. But cows with longer winter coats will get filthy if they aren't given plenty of bedding and an occasional clipping and brushing around the hind quarters.

"Come bossie . . . ."

The cud chewers stop. A head turns, then another, sensitive ears dished in the direction of the familiar call. An old cow with sagging udder jacks to her feet. She rises hind-end-first, momentarily "kneeling" on front wrists doubled under in a way that would break our own. Heaving back, she straightens her forelegs each in deliberate turn.

Arching her back, tail cocked, she urinates. The cows that were lying down are all up now, all stretching and urinating or defecating, yet likely retaining some measure of appreciation for the clean barn floor.

Herd Instincts

Instincts rooted in survival keep a herd of cattle together in all things. A fly with its compound eyes is hard to swat. Herds, flocks and schools borrow from the fly. Watch a cat stalking a flock of sparrows. Though busily feeding, the flock is ever alert to danger and protected by its own compound eyes and ears.

Within a herd there is a pecking order, just as there is in a flock of chickens where the system of hierarchy was first described. Normally a mature bull would be at the top of the order. But take the bull away and a low-key test of wills begins with pushing and head bunting until one cow proves stronger or more persistent than the others. In

managing large herds of cows, enough feeding and watering spaces have to be available to make sure cows low on the pecking order aren't deprived of their needs.

Unless a cow is seriously ill, about to calve or in strong heat, she will stick with the rest of the herd. One cow alone, without calf or others of her kind, will strike up a friendship with a horse or whatever other large animal may be pastured with her. Being social, and more curious than any cat I've ever known, a single cow will follow her keepers anywhere a door or gate is left open.

Size

Dairy cows are large—800 to 1700 pounds and sometimes larger with individual Holsteins or Brown Swiss. But it takes time. They are slow maturing, compared to some of the smaller farm animals. Though they may calve at the age of two, it is unlikely they'll reach full structural size for another six months to a year. Some poorly-fed cows may go to four-and-a-half years before reaching this "structural" maturity. All well-fed cows can be expected to show an annual gain of fat and flesh right on through their prime years—ages six to nine.

Walking

Dairy cows are angular. They look ungainly. They plod along with such resignation and acceptance of a body God shared only with a moose. As if that weren't bad enough, along came people imposing pendulous udders to slap back and forth with each step. Could they move any faster if they wanted to?

They can and they do, especially when they are young. They run, gallop, rear and flail with their back legs. A young heifer or bull is a laughing, joyful sight charging around a pasture. Sometimes they break into a graceful trot, the more usual fast gait for a full-grown animal, with tail looped long behind.

Sometimes young animals will break into a chase for the apparent hell of it. More often something sparks them off—being turned into a new pasture, or one getting stung by a bee. Bringing a new cow into the yard or taking one away can send the rest of the animals into an uproar, complete with anguished bawling.

Usually all this is great fun to watch, until a cow heads for a fence at a dead run and then . . . she's . . . Damn! For cows *can* jump. Hey diddle diddle. Again this is especially true for young ones. They can jump or trample most common fences.

Our dairy cows can swim, too, but regardless of special talents they are best at walking, it's true, and then on ground that gives beneath their split or cloven hooves (corresponding to our third and fourth fingers or toes). Sod ground, turf or soft gravel are their choices, and when given the chance they will go out of their way to avoid deep mud, loose rocks or any slick, hard surface.

A hungry cow may charge for the barn when she's called. Don't let her steam on into the barn, though, because there is too great a risk she'll slip and fall when she reaches the concrete or wooden floor.

If cows are out of water or food or they're late being milked, they let the world know it with an insistent mooing. (I have read that Brahmans don't moo—that they grunt. This has me wondering if a Brangus munts or groos.) The mooing is loud but not so pained as the bellowing when they are upset or in heat. At the other extreme comes the low, warm rumble of a cow reassuring her calf.

When cows are regularly fed and milked they soon learn where they belong. They will file quietly in and stand waiting for their stanchions to be closed or chains snapped. Occasionally, out of orneriness or forgetfulness, a cow will step into the wrong stall. In a barn with many cows this can set up a chain reaction that would try the patience of a turtle. At a time like that it's hard not to believe cows have wicked senses of humor, as they back and shove and stumble about, thoroughly manuring everything in sight. Screaming at them and whacking about with a cane only makes the situation worse.

Regardless of the quality of feed in the pasture, cows will always relish a taste of grains when they come into the barn. They'll chew into a pile of grain like a kid in a pie-eating contest. Then, as the pile dwindles, that long tongue comes into play, sweeping the manger floor for the last crumbles.

After graining and milking the cows are released. It is a quiet time, with all sense of urgency gone. The cows slowly back from their stalls, turn and amble to the door, nosing a window, tentatively licking the handle of a hoe standing in the corner; it's a dreamy time. "Shoo boss! . . . Hye!"

Returned to pasture, they will graze for a while, then slowly settle, one after the other, finding a level spot to lie down, front-feet-first, and to chew: 34,102 . . . 34,103 . . . 34,104. . . .

CHAPTER 4

Buying A Cow

Buying a cow is like buying a used car—90 percent of the gamble is in the person you buy from. Your best bet is on an honest neighbor who would like to see you and the cow succeed. Yet even the most sincere and honest person can't know certain things about a cow without the help of a veterinarian.

The prettiest cow with the best past record of kindness and performance is useless if:

(1) Her udder and teats aren't in good order.
(2) She is not carrying a calf.
(3) She has not been tested and certified free of Bovine TB and brucellosis, preferably within the past month. (Johanes disease too, if this is a problem in your area.)

From a Neighbor

Under ideal conditions I would find a cow for sale down the road, from a good neighbor, a sound and healthy-looking cow. And if the cow had not been recently checked for TB, brucellosis or pregnancy, I would ask if the neighbor would mind my calling the vet to have these things done. I would pay the vet bill.

Checking the teats and udder and finding out if the cow is pregnant will take only a couple of minutes of the vet's time. In some cases,

especially in heifers, an experienced hand can feel the lump of a developing fetus that is only seven weeks old. It is done by reaching
into the cow's rectum and feeling the uterus through the thin-walled
large intestine. In older cows the fetus may have to be two months
along or more before it can be told from other lumps or scar tissue left
from previous calvings.

There are laws in the U.S. and Canada about regular testing for
TB and brucellosis. But it doesn't always get done. I have bought cows
that hadn't been tested for well over a year. It was stupid, but I lis-
tened to the wrong people who said not to worry. "Hasn't been a case
of brucellosis in ten years. Don't remember when we had the last case
of TB."

Well, Vermont just discovered a lot of brucellosis and several
cases of undulant fever, the human form of this bacterial infection. It
was quite a shock. If a cow reacts positively to either the TB or brucellosis tests she has to be destroyed. If we contract either of these
diseases we could be in for months or years of medication. I'd rather
pay the $10 or $20 and wait for a clean bill of health. The TB skin
test takes three days to read. The blood test for brucellosis takes 10 to 15.

Looking Elsewhere

If the friendly, honest and obliging neighbor hasn't got a cow
for sale, it's time to get serious about knowing cattle and markets. Go
shopping, but leave your money home. Get to know cows by looking
and asking and comparing. Get to know the prices of dairy cattle and
different grades of beef animals.

Before hitting the open market, try to find a cow through word-
of-mouth leads. Your veterinarian will want to see you started right, or
should. The artificial breeder may be able to help. The local agricultural
extension service representative, feed and farm supply dealers and
drivers for the local livestock auction barn are other people to ask.
Then check the local newspapers for farm auctions or cows for sale in
the classified ads.

Visit dairy farms. The more modern and competitive farms are
culling low-producing cows like never before. In spite of excellent
breeding, a farmer may have a young cow that is not producing enough

milk to pay her way. After four or five weeks of milking the farmer can tell pretty well what the cow will produce over the next eight or nine months and will know that after so many months there will be no sense keeping her in the barn.

A Jersey headed for 4000 or 5000 pounds of milk or a Holstein headed for 8000. These cows would break a commercial farm but they would be ideal producers for a family. If there is a farmer willing to talk about it, ask about buying one of those cows when the time comes that she ordinarily would be shipped to the market. Agree on a price plus the cost of having her bred. Done.

The Jersey cow. 800 to 1200 pounds. Light fawn to black, though the deer-like color is most commonly associated with the breed. They may have white patches and the tail switch may or may not be black. Nose, black, ringed by a light muzzle. As a breed, Jerseys are lowest in yield but highest in production of butterfat (5.1 percent average) and total solids (upwards of 15 percent). They originated on the Island of Jersey in the English Channel, and the first registry association was founded in the United States in 1868. Jerseys are the smallest dairy breed.

Too Much Milk?

If you buy a cow that, when she comes into milk, gets to the point sometime in the second month of lactation where she's producing whopping quantities of milk (say around 50 pounds daily for the small breeds and in the neighborhood of 60 pounds for the large) even though you're feeding her only moderate quantities of food, maybe you have more of a cow than you want. She could be a cow whose genetic

The Holstein-Friesian cow. 1200 to 1400 pounds. Black and white, though to qualify for registration the association will not allow there to be any black in the tail switch or below the knees or hocks. The association recently has begun registering a variation that is red and white. Herd averages of 16,000 pounds of milk a year per cow are common. However, Holsteins give the lowest ratio of butterfat, averaging 3.5 percent. The breed's stock came from Holland and the present U. S. association was founded in 1871 out of an amalgamation of the Breeders of Thoroughbred Holstein Cattle and the Dutch Friesian Associations.

potential for production approaches tops for her breed. She may be trying desperately to meet those internal demands the only way that she can under moderate feeding which is to draw heavily on her own bodily reserves of fat, minerals and protein to produce those quantities of milk. (See Chapters 11 and 13 for more information on feeding and milk production.)

Cattle Auctions

The countryside is dotted with regularly-scheduled livestock auctions or commission sales. Some fine animals that would fill every need in a family cow go through these sales. But spotting one takes more experience. For every good cow there are likely to be dozens of clunkers. Nothing is guaranteed or even vaguely promised about the health and past performance of the cows sold here.

If you do go to a commission sale, go early. Most of the animals arrive before sale time and are distributed and sorted out, depending on how fancy the operation is. Walk around the pens and don't hesitate to ask questions.

With any luck there will be a chance to look a cow over closely before she goes on the block. It may be possible to find out where she came from and why she is being sold. It could be that she is a perfectly sound and able cow—just one that wasn't producing enough milk to pay her way on that commercial farm.

It wouldn't hurt to visit the auction a few times to find out how it works. Maybe buy a small pig or a lamb so that when the time comes and you feel like bidding on a cow it won't be an entirely new and awkward business.

These commission sales are also among the easiest places to sell a cow when the time comes. Often reserve bidding is allowed, which means that for a small fee the cow won't be sold for less than your set price. If no one bids higher than the reserve you take your cow home and try for a better day at the sale.

Consignment sales of registered cattle are sponsored by the different pure breed associations. Any farm with a sign out front advertising registered cows ought to know about upcoming sales for their breed. The cows at these sales usually will be bred heifers about to calve (springers). They will be guaranteed free of TB or brucellosis. They will be expensive.

What to Look For

The best family cow for beginners will be due to calve, or freshen, with her second or later calf within a couple or three months. She'll be at the tail end of her last lactation, be easier to milk, and is not likely to be bothered about not being milked out completely through your first week of learning how to milk.

SIZE

She'll be a small cow, likely—up to a thousand pounds. And most people seem to prefer a family cow that's heavy on the Jersey, Guernsey or Canadian side. She'll give less milk than will a Holstein or Brown Swiss, but what she gives usually will be higher in butterfat and total milk solids.

PRODUCTION: MEASURING MILK AND BUTTERFAT

A farmer whose cows are on official test—meaning someone comes in once a month from the Dairy Herd Improvement Association (DHIA) or the like and accurately measures each animal's milk and butterfat production—will have production records that stand up to questioning in hard-nosed cattle dealing. Records from another system called "owner sampler"—where the farmer weighs each cow's milk once a month and sends a sample in to the DHIA or whatever lab for a fat test—is not worth much in the market place.

Measuring the "cream line" on a bottle of milk will not give an accurate reading of butterfat production, because the fat droplets come in different sizes and the smaller the droplets the tighter they will pack together, giving a picture of less cream. (See Chapter 13.) Age, breed and stage of lactation all affect the size and proportions of big to little fat droplets in a cow's milk. Taking all of these into consideration, you could conceivably come up with a "cream line" that indicated a whopping 15 or 20 percent fat production for a young Jersey or Guernsey early in lactation.

THE VARIATION IN THE FAT CONTENT
OF MILK OF INDIVIDUALS OF THE
DAIRY BREEDS

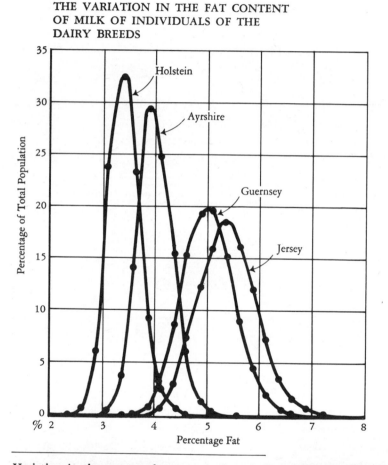

Variation in the average fat content of the milk of individuals of the dairy breeds. Only Advanced Registry records over 306 days in length were used. Included were 2,665 Ayrshire records averaging 4.03 percent, 36,861 Guernsey records averaging 5.05 percent, 26,773 Holstein records averaging 3.41 percent, and 31,297 Jersey records averaging 5.41 percent. The higher the curve the less is the variation from the average. The fat content of the milk will not alone characterize the breed, as some cows of the four breeds have the same average yearly fat test.

(From Missouri Agricultural Experiment Station Bulletin 365.)

AVERAGE COMPOSITION OF MILK
FROM FIVE BREEDS OF COWS

BREED	Butterfat (%)	Total Solids (%)	Solids Not Fat (%)
Holstein	3.45	12.29	8.84
Ayrshire	3.85	12.98	9.13
Brown Swiss	3.91	13.28	9.37
Guernsey	4.92	14.20	9.22
Jersey	5.14	14.90	9.76

(From USDA *Yearbook of Agriculture*, 1939.)

A cow's overall production potential involves both milk and butterfat, either of which varies tremendously between breeds and even between individuals of the same breed. The only way to compare two cows is to see what they are doing in terms of some common denominator. The system used today calls for multiplying a cow's milk production by 0.4, her fat production by 15, and adding the products to give a "Fat Corrected Milk" (FCM) reading that says what she's doing in terms of so many pounds of 4-percent-fat milk.

The only reason for talking about FCM is so that the term doesn't come as a curve ball from some pro. FCM, DHIA. . . . Are they all that important or necessary with the family cow? I don't think so. I think most of us will choose cows on the basis of looks and on whatever we can find out about breed. Knowing breed capabilities and averages is about all it takes to select a cow that will tend to produce a certain quality of milk.

BODY CONFORMATION

No one needs a ribbon winner for a family cow, but some of the points on conformation that judges look for in a show are good indicators of strength and durability. Faulty lines often are the kinds of twists, turns and sags that can be expected in a 10- or 15-year-old cow but that shouldn't be there when she's in her prime.

The milk cow's head and neck should be slim with a top line bowing slightly to the shoulders. From there the top line should extend almost on a level to the tail. A line from hip to pin bones should be about level, but it's all right if this line slopes a bit down to the rear. Looking at the cow from front or rear, her legs and feet should not be splayed, knocked or crooked. She should not be "winged," meaning that her shoulder blades stick way out from her body. She should not be "sickle hocked." This is a defect of the back legs where a permanent bend (seen from the side) in bone or joint places her feet so far forward that the inside curve of the leg looks like the blade of a sickle.

She should have strong pasterns so that she is up on her toes. An appearance of weak pasterns may be the result of untrimmed hooves.

Some of the lines you don't want in a young cow because they foretell early aging or weaknesses are seen in this venerable Jersey. Notice particularly her swayed back, pendulous udder and winged shoulders. You'd have trouble getting a milk pail under this cow—unless perhaps you could teach her to stand with her hind legs on a pedestal. Perhaps you could, in which case you might find you've wangled your way into a good cow for cheap because they didn't have the time to bother with her any longer on a commercial milk farm. For all the rough points there's nothing sickle-hocked about this old cow. Judges like to see this cow's tendency toward a straight line touching pin bone, hock and dew claws.

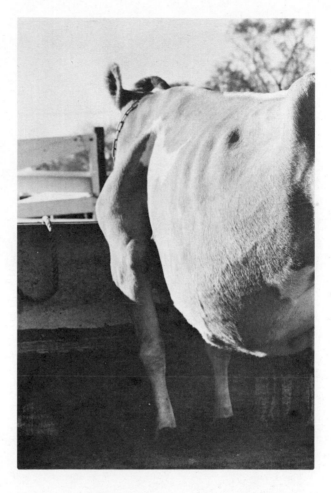

Winged shoulders.

But then this could be a chicken-or-the-egg question, because if the pasterns were up to par the cow's toes would more likely get the necessary wear and tear and wouldn't need trimming.

The cow must be able to move around smoothly and easily without limping. Her joints shouldn't be swollen. Check for swelling above the hooves. A circle of puffiness there may be a sign of foot rot or of nutritional problems.

Some older cows come down with a chronic condition known as "crampy," where they awkwardly lift and shake a back leg. It's something like arthritis and isn't considered a serious problem. But, being an amateur, I would want a doctor's opinion before buying a cow with any kind of limp.

To eat, the cow has got to have a good set of incisors at least.

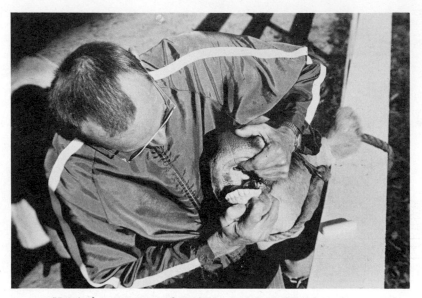

Here's how to get at those lower incisors when you want to estimate a cow's age. There's no danger of being injured through biting since the cow has no upper incisors. There's also a wide gap between her lower incisors and first molars —room for several fingers and a firm grip.

And the jaws should meet squarely, meaning the lower jaw can't be under- or overshot. By counting the incisors and checking their wear it is possible to get some idea of a cow's age.

OUTWARD APPEARANCE

A cow should look alert and clear-eyed. Her muzzle should be moist. There shouldn't be discharges from her eyes or nose or any persistent coughing. She should be responsive to what's going on around her. Her hair should have a gloss, be quite smooth, and lie flat except perhaps in the case of a cow wintered out or in a cold barn, whose coat may be fairly long and shaggy. It's a healthy sign for a cow to be eating or chewing her cud.

The skin should be thin, pliable and elastic so that if you pull away a loose pinch of hide and let go it will snap back in place. A sick

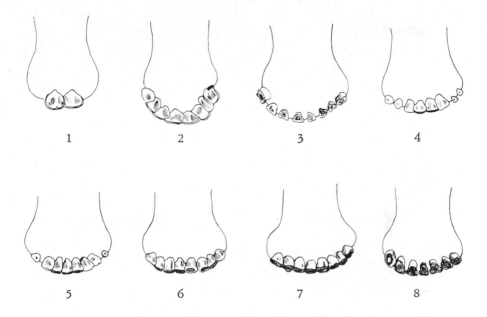

Numbers and condition of lower incisors give clues to a cow's age. (1) Birth. Two "milk teeth" are almost always present. (2) One month. Full complement of eight baby incisors will usually be present. From this time through about 18 months there will be no change in the number of incisors—only increasing wear. (3) About 1½ years. Baby incisors have become very worn and appear to be spreading apart. (4) Two to three years. Baby incisors are being replaced from the center out by permanent nippers. By the age of three the inside pair of permanent incisors will be showing indications of wear. (5) Three to four years. All but the last pair of baby incisors have been replaced by permanent ones. (6) Four to five years. At last a full complement of permanent incisors! But notice how the most recently installed (outer) pair are not yet worn. (7) Five to ten years (about). Telling the precise age of a cow through these years will be a guessing game. She has all of her permanent incisors. All are somewhat worn, but not badly. The incisors are tight together. (8) Twelve years and on. An old cow. Her incisors are becoming very worn and isolated as the gums recede. When she begins to lose incisors, she is said to have a "broken" mouth; this is a drawback because it impairs her ability to graze.

cow's hide may sag back slowly and even show the pinch mark for a moment or two, a sign of serious dehydration.

The cow should have a rounded, barrel-like body—so rounded that to the inexperienced she probably looks eternally very pregnant. Her ribs can show. They should spring away from the backbone so that while from the point of her back down she is peaked or sloping, it still is not like a church steeple.

UDDER AND TEATS

The cow's udder should look as if it came with the old girl, not as if it resulted from some scientist's experiments in organ transplant. It needn't be a huge udder. There's not much correlation between udder size and milk production. A large udder could be filled with fat or scar tissue that's totally nonproductive. Neither is there much correlation between large "milk veins" and production, though these veins on either side of the cow's underbelly should be about equal size.

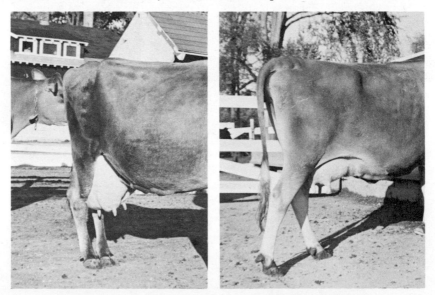

Large udders don't necessarily mean more milk. The young cows carrying these udders are within a year of each other in age and are on a par so far as production is concerned. Yet the cow on the left is headed for an early retirement due to weak udder suspension. The cow on the right has an udder that's likely to hold up for several more years.

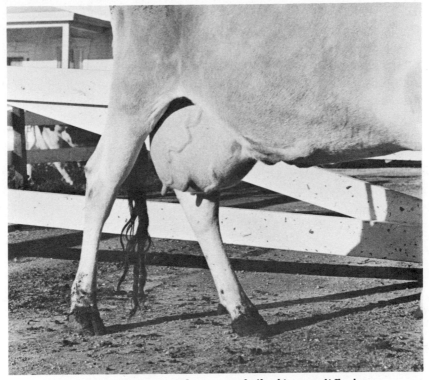

Poor teat placement makes a cow built this way difficult to milk. Imagine how she would look just freshened. A milker would need long arms to reach the far teats.

The udder should have four teats evenly spaced and all of a size that fit comfortably in a loosely-closed fist. Fat teats are often meaty and hard to milk. Short teatlets are impossible.

The teats on a heifer or any young cow just freshened (in milk following calving) may be on the sort side. Often they will lengthen with lactation or age.

The bottom of the udder should be flat rather than bulbous, and be well attached at both the front and rear. Looking from behind, the skin of the udder should join the cow's body in a broad band from thigh to thigh. If from any angle the udder looks like a water-filled balloon it means that the muscles and ligaments intended to hold the udder against the abdomen are breaking away. Generally speaking a narrowly-attached udder is one that will sooner break away, leading to a pendulous udder and teats so near to the ground that they are difficult to milk and are more likely to get injured. There should not be deep valleys between the four quarters of the udder, but a shallow valley between right and left sides of the udder as seen from behind does indicate good suspension.

The udder should be pliable and should deflate considerably after milking, leaving folds of skin as seen from behind, and no hard areas. Make sure that the udder and all the teats are working properly, by taking a finger and thumb squirt from each and looking to see that the milk is not clotted or off-color.

Sometimes, through injury or disease, a cow springs a side leak in one of her teats, making it difficult or impossible to milk her by machine. The leak may be no problem in hand milking if it is where it will be covered by the palm.

Sometimes a part of a cow's udder gives up the ghost, leaving one or more teats dried up for good. A cow can go on for years giving almost as much milk as before because the remaining quarters compensate by increasing their capacity.

But a veterinarian warned me against buying a cow with a leaky teat or a shot quarter if it isn't known how they came about. They may have resulted from an acute infection, and the theory is that a cow that became that seriously infected in the past is more apt to come down with another bad bout.

HORNS

If there is a choice between a cow with horns or one without—either because she is *polled* (genetically hornless) or because the horns were removed when she was a calf—I'd say take the one without. Cows with horns may be more beautiful, but they get into more trouble. They can hurt people, not intentionally, but it hurts all the same. The horns can be used to bully other cows, and they get used for all sorts of mischief like breaking windows, roughing up barn siding and hauling clothes off the clothes line. They can get wedged and broken off, leaving a bloody mess and maybe even necessitating a visit from the vet to finish the job properly.

TEMPERAMENT

The cow's temperament is important. There are opinions about different breeds: This breed is docile, the other nervous. Actually there

is much more difference from one individual cow to the next, regardless of breed. Everyone thinks of Brahmans as being the ferocious
cowboy-manglers. But if they are raised properly the cows are wonderfully gentle. Some breeds are prized as riding animals in parts of the
Orient, known for their easy gait and ability to travel 50 miles or more
a day.

No one wants a "kicky" cow. They are a nuisance at best, and
extremely dangerous around knee-high children. A nervous cow may
have been born high strung. Maybe she got that way through mis-
treatment. A heifer coming up to her first lactation, never having been
handled or had her udder massaged, is bound to be a bit nervous.
She'll lift her back legs and shift about some. I wouldn't say that here
is the sign of a troublesome cow. But if she really rams about and lashes
out with her back legs, pass her up in favor of a cow that won't keep
you guessing.

*The cow with horns can get into more trouble, like rooting
up barn siding or hauling clothes off the clothes line. Yet
horns are preferred by some, for their looks or else for the
fact that they provide a solid hand or tie-hold, a possible
big advantage on a flighty young animal.*

There are times when the quietest cow will kick, such as when she has a sore teat. But there are ways to avoid those kicks as well as ways to restrain the most devilishly persistent kickers. These are gone into in Chapter 5 on handling cows.

Prices

Milk cows aren't cheap. The price of a good dairy animal will usually lead local beef prices by five, ten or more cents a pound, live weight. I have known people who would never pay more than the local beef price for any cow. That way they know if she doesn't work out for milk they can get their money back in the meat market.

I also have made a farmer quite mad by pulling out a girthing tape to find the weight of a cow he was selling. I went ahead and measured her anyway. I might willingly have paid above the going price for beef, but I wanted to know the above-beef risk I was getting into.

To estimate a cow's weight, measure the diameter of her body just behind her front legs (heart girth) using a girthing tape or chain. Or use a standard measuring tape and convert the inches to pounds using the table in the appendix. The cow should be standing straight-legged and on the level, and the tape should be pulled snug but not tight.

The most expensive cows are the pure-bred beasts registered with one of the breed associations. In most cases all registration guarantees is that the cow came out of registered parents of the same breed and that she conforms in color and pattern of marking to that particular brand of cow. This is not a whole lot to go on, it would seem, but years of controlled breeding and culling have reduced the number of surprises and possible disappointments that are more apt to be lurking beneath the skins of unnamed, unregistered cows.

When I plop down money for a Jersey-like cow, content with the knowledge that she's going to give Jersey-like milk, I really should send a note of thanks and a couple of bucks to the Jersey Association. Where would I be without them?

CHAPTER 5

Handling
and Grooming Cows

Patience is *the* ingredient for handling cows. It's more important than cleverness, force, or will. With patience strange things are possible. Cows can be taught to climb stairs. They can be taught to climb on a revolving carousel milking platform. Neither stunt will help the family cow, it's true. But the thing is that without patience cows become kicky, nervous creatures that do everything wrong.

Once cows have learned some simple tricks within a simple routine, half the work of caring for them will have been swept aside. Only the occasional problem may come up when the routine is or has to be broken.

Bringing Them Home

Cows that come when they are called are a great time-saver. They will come to any name or noise once they have learned to associate the sound with something to eat. In breaking in a new cow or heifer you may have to take the bait to wherever she's lurking in the pasture. Call her repeatedly and give her a taste before slowly heading back for the barn. The first couple of days several tastes may be needed to coax her all the way home.

Sometimes cows that are well fed and not achingly full of milk won't come when they're called, preferring instead to lie and ruminate on some distant knoll. But which distant knoll, and how to find them?

47

With patience cows can be trained to do many things.

Cow bells were an answer in the old days, being a great help in finding cows that often were allowed to range free. They could have had other functions as well. In England they are sometimes used to alter the pecking order in a herd. I've been told that in some cases all it takes is one or a larger (louder) bell around the neck of a bullied and miserable cow to lift her back into contention for the feeding and watering troughs.

On the other hand a cow that has never worn or heard a cow bell may be terrified by the constant bing-bong. If you want to use a bell it might be best to start out with a small, or at least a quiet one.

Even if a cow is not wearing a bell it won't be as difficult as it might seem to find her on the largest pasture, unless maybe she is hiding out with a newborn calf. Much of the pasture may not normally be used. What she does use will probably be divided by habit into places she prefers for grazing or for bedding down. Often there will be established trails linking these areas, and after a while a person will be able to guess pretty well where the cow is likely to be at any time of the day.

Time that may be lost searching for a cow won't be made up by running her home. She can run faster than you, and once on the run there's no telling where she'll go. One thing is certain. She won't be in the mood for milking when she gets there.

Herding and Driving

When herding a cow or cows, keep in mind that they have this animal instinct scientists call "flight distance," some number of feet they like to keep between themselves and any threat. The distance varies with the individual cow and the nature of the threat. Is the threat moving? How fast? Does it carry a stick or something flapping? Is it noisy?

Anything loud and flapping may upset the cow and put her to flight at 30 feet or more. There's seldom any need for that. It's better to approach a cow slowly, get her turned in the direction she's to go, then give her a gentle encouragement with a wave or a hand clap. If she doesn't move, give her a swat on the rump. Increase the pressure slowly. A handful of pebbles or the slenderest twig make wonderful allies for moving cows.

Herding cows that have got out of the pasture can be more difficult because they'll be keyed up and easily spooked. There's nothing impossible about the situation, though. For one thing the cows are on unfamiliar territory. With gentle crowding they're more apt to go back the way they came than to push on into the unknown. Given a chance they will often work their way back to that hole in the fence where they escaped—a hole it might otherwise take hours to find.

The hardest part of driving cows is in making them turn at a gate, intersection or through that break in the fence. With two people, bring the cows to the fence line somewhere above or below where they are supposed to turn. As soon as they are moving along the fence line one person should circle far around to take a stand fifteen or twenty feet beyond where the cow(s) should turn. A mistake that is often made by people who don't know cows is to stand too close to the turn or opening in the fence—right at the turn, in fact, so that the cows being herded from behind take fright and scatter before they reach the turn.

Another way to corral a bunch of cows is to catch the quietest one and lead her home, hoping that the others will follow.

On any pasture where cows are allowed to roam free for any length of time it helps to keep them used to you and the idea of coming when called. This is easily done by once-a-week visits with a little grain —or salt if they haven't been left with a continuing supply of this supplement.

When heading cows, a person should stand well back from the fence break or gate where it's hoped the animals (being driven from behind) will turn. Farmer Frizzle, above, is standing just about right.

A small, easily sealed-off enclosure near a pasture gate may be useful as a corral for catching cows—particularly young stock—that may have become difficult to manage after a summer of running free. Often the mere fact that they find themselves suddenly confined to a much smaller space is all it takes to bring previously trained cows under enough control to be haltered and led. You won't get off that easily with anything but calves unless the animals were previously trained to ropes and halters. For this reason it is important to get cows used to close contact and handling early in life.

Leading with Rope or Halter

A quiet, well-trained animal can be led by a rope or collar around her neck. But a halter will give better control and is a necessity for cows that are a little balky or that are being taken into any unusual

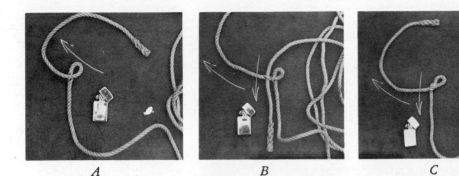

A   *B*   *C*

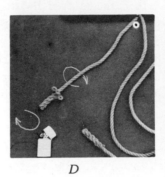 

D   *E*

Ten or more feet of stout, three-strand rope makes a halter to fit any size calf or cow if you twist, turn and loop it as shown in these photographs. Before shaping the halter both ends of the rope should be back-spliced, taped or whipped to prevent fraying. (1) Beginning eight to ten inches from one end, open the rope coils and feed the short end through, forming a loop that's only a hair larger in diameter than the rope itself. I'll call it Loop "A." (2) Open the coils once more, on the short end side of Loop "A" and feed the long end of the rope through. This locks Loop "A" in place. (3) Grasp the short end of the rope firmly in both hands and untwist until the braid opens in a clover-leaf or looped strands. (4) Feed the long end of the rope through the three loops and pull up to form the halter's nosepiece. (5) Now run the long end of the rope around and through Loop "A." Pull up to form the halter's headpiece. Both nose and head piece are adjustable. To halter a calf or cow, first snare the animal's nose, then run the larger loop over the back of the head and below the ears (and horns, too, if she has them).

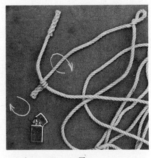

F

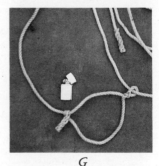

G

51

situation. For extremely uncooperative cows a nose lead is sometimes used. This instrument puts pressure on the sensitive cartilage separating the cow's nostrils by degrees, depending on how hard it is pulled.

A cow can see to run only if her head is down, which is one reason why a collar or neck rope won't control a stubborn animal. Also, a nose lead or halter gives the handler the greatest leverage for turning the cow's head because it pulls from the end of her long muzzle.

For the show ring and by custom, the way to lead a cow is from the animal's left (near) side, by her head, and with the lead rope in the left hand. From there the cow's right is called her "off" side.

If a cow is stubborn and doesn't want to move, it won't do any good to stand in front and pull. Stand beside the cow, coaxing her forward with the lead while giving her a slap on the rump with your free hand. Use a stick or switch to tap her rump, or reach back and pinch the base of her tail. Or pick up the end of her tail and use that as sort of a second lead.

Don't yank a cow's tail, and never kink or twist her tail. It is certainly painful and it may make them move. But I have seen tails broken on panicked cows that even then would not budge.

If the cow tries to break away from you, pull her nose up in the air and toward you. This forces her to turn. Round and round you may go, but as long as you keep her head high and turned, the odds of winning are in your favor. Don't ever tie or twist the lead rope around your hand or wrist. You have to be able to let go fast if you lose control.

Loading a Cow

In loading a cow on a truck the end of the ramp or tailgate may be a couple of feet off the ground. But the lower the step and the more gradual the incline of the ramp the better. If you haven't a ramp, look for a hill to back against or drop the back wheels of the truck into a ditch.

It's best to have three people to load an ignorant cow: one to pull and the others to push, prod and lift. Sometimes when there's a step involved it helps to pick the cow's front feet up, one at a time, placing them on the ramp or tailgate. Two people can give the cow a great boost by locking their wrists behind her tail for a lift and shove.

About this time the really obstinate animal will collapse. It's a tremendous test of patience, and a time when I've seen cows brutalized by people I'd never have thought capable of raising their voices in anger. Ideally, the thing to do when a cow lies down is walk away. Have a beer and let everyone relax. Barring that approach, there is a way to get a cow up that doesn't do her any physical harm. This is to cup your hands over her nostrils until, in a desperate attempt to get a breath of air, she struggles to her feet.

Traveling in Style

Don't carry a cow on a truck that has a steel bed. Bolt down a sheet of plywood or chip board to give her some traction. There should be sideboards going at least four feet above the bed of the truck. Solid panels all around are best, especially if the cow has horns that could get jammed and broken in a hole or crack between slatted sides.

A cow should be tied short in a truck, left enough freedom to lie down but not enough to get her nose above the head or sideboards. If she can get her nose up, a feisty animal may try to jump out.

No need hiring a truck to haul the new cow home. A simple trailer suits this Bossie to a "T." (Courtesy UVM Extension Service.)

A little bedding of straw, dirt or sawdust under the cow's back feet will give her better traction and help to keep her clean.

There may be no need for a ramp or hill to get a young cow off a small truck. She'll gladly jump, and with ease.

Jumping

Speaking of jumping, calves and heifers can get to be incredible escape artists. A remedy for those that jump and climb through fences is a neck yoke affair called a *poke*. We had one heifer who refused to stay in a new summer pasture even though she was with friends, had plenty of food and running water. She wanted to be by the barn where she'd spent the winter. A fancy triangular poke around her neck didn't help. New posts and wire didn't help. What finally did work was tethering her inside the new pasture for three days round the clock. By then she was used to her new surroundings.

Handling cows has its few hazards—not because cows are malicious animals, but because they are large and we are slow. We're often unaware of possible danger in a carelessly hooking horn or even the casual nod of a hornless head that nonetheless has all the clout of a 30-pound sledge on a rubber handle.

Kicking

Cows stomp on people's feet with their front hooves and sometimes kick with their back. I think they mostly do the stomping for laughs because once they have planted a hoof on a soft toe, shoving and shouting only seems to bring more pressure to bear. I get my own laughs with steel-toed boots.

A frolicking heifer may lash out with both hind feet at once. A cow in a stall uses one leg at a time, dealing either of two basic kicks. There's the quick jab from the floor, directly back, and the swipe that begins with the leg cocking forward, then scything out and back.

To avoid getting kicked, try first of all to buy a quiet, kind cow. This shouldn't be hard. The opposite is the exception. But it is reason-

A cow can't kick when her tail is hiked in the air as shown. Avoid a common mistake, which is to grasp the tail farther toward its tip where it is weaker and may possibly break if too much pressure is applied.

able to ask point blank if the cow you are thinking of buying is kind.

Make it a practice not to walk directly behind a cow. Give them two or three feet—space enough for you to see the foot coming. After you and the cow have come to know each other you probably will never have to worry. You'll be able to lean your back to her tail without the slightest fear of getting a lift.

But warn strangers, and especially children who often become excited and noisy, crowding about and having no idea that the gentle cow you're milking could become fed up with the racket behind her.

Even with a cow you know and trust it is a good idea never to sneak up on her. With noise and talking let her know where you are.

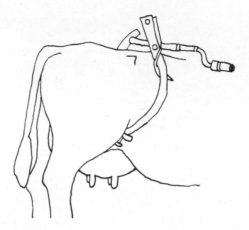

One of several commercial devices for restraining a kicky cow.

When walking around and behind a cow put your hand on her back so that there is no doubt that she knows you are there.

The scything kicks are most often brought on at the start of milking. The udder or teat may be bruised, swollen or nicked. It is not so much a vicious kick as it is the cow's only way of saying, "Hey, that hurts!" To avoid that kick when milking, tuck in close with your left shoulder pressed into the cow's flank. The forward part of that kick usually is weak. It can do no more than push you out of the way.

When having to stand directly behind a nervous cow—say in order to take her temperature—lift her tail up. It is impossible for her to kick when her tail is firmly held this way.

A confirmed kicker can be restrained during milking with a rope hobble or tee tied around her hocks or with one of the stop-kick devices sold in stores. Or she can be sold.

The Contented Cow

GROOMING

Trust and affection are better than restraints of any kind. They come through grooming, and grooming pays. There were some tests conducted in the early part of this century which showed that cows that were regularly groomed gave up to 8 percent more milk. About that time one of the leading dairymen in the United States, Professor T. L. Haecker of Minnesota, wrote, "If you so handle cows that they are fond of you, you have learned one of the most important lessons that lead to profitable dairying. A cow's affection for the calf prompts her desire to give it milk (and) if you gain her affection she will desire to give you milk."

Fine. But how does a person go about gaining the affection of a bony, hulking cow?

First, remember those large and excellent ears and keep the noise down. Think of the cow as an anxious mother, which she is, and move quietly and slowly around her. Avoid sudden moves and sudden clangs and bangs.

Second, the cow loves attention. She loves to be scratched and brushed and talked to. She's got pleasure points for scratching—just behind her horns or pole is a favorite spot, or the base of her tail. She loves being scratched and rubbed under her chin and neck. The brisket is another sensitive area, especially when the cow is in heat.

A lone cow will strike attachments to other animals or even things. A cow and horse pastured together may form a close bond. One of our cows fell in love with a rubber tire swing hanging in a tree. She'd stand neck and shoulder to that tire for hours as it gently swayed in the wind.

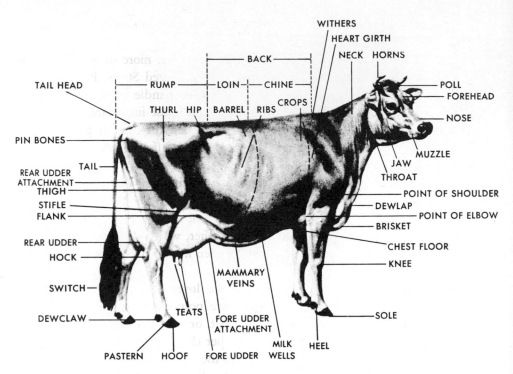

Parts of a dairy cow. (Courtesy, The American Jersey Cattle Club)

The more and closer attention a cow gets, the sooner you will be likely to spot any health problems. There might be a cyst or a small cut that could get infected. Or it might sooner be noticed that the cow is getting thin or her coat is getting scruffy and dry—maybe the first signs of a nutritional problem. Clipping and brushing cows, particularly around the udder and flanks during the long-coated winter season, guarantees a more comfortable cow and cleaner milk.

HOOVES

Cows' hooves grow continually, like fingernails. If they don't get filed in the normal course of pasturing you should file them off periodically with a heavy-duty rasp. Hooves that become badly overgrown may have to be cut back with hoof trimmers or with a hammer and chisel. As in all other handling procedures it is much the best to get cows used to having their feet looked at and worked on at an early age—six months or less. Get a calf used to having her feet picked up, at least for examination. That way it will be an easy routine when she gets fully grown.

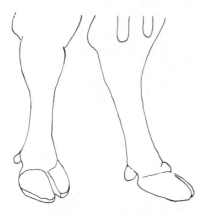

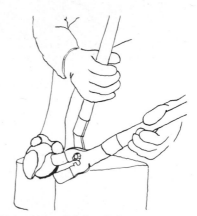

Before and after trimming hooves that became seriously overgrown.

Trimming with hoof shears, rasp or chisel will prevent lameness in cows.

A cow on pasture will usually manage to trim her own hooves in the course of cruising for food and water. In the event the hooves do begin to grow ahead of natural wear it would be best to file them back at regular intervals using a coarse rasp. If hooves grow ahead of rasping, it will be necessary to use hoof shears or a wide chisel and mallet to trim them back. Until you and the cow are used to the operation, go slowly, taking only a little excess hoof off at a time. Cut back too far and you will draw blood as happens when a fingernail gets ripped back to the quick. A large animal may balk at having her feet worked on the first time around. Don't hesitate to call in a veterinarian or professional hoof trimmer.

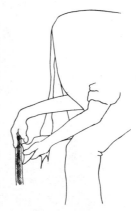

A periodic rasping may be all the hooves need.

A wide wood chisel works for trimming if you haven't got shears.

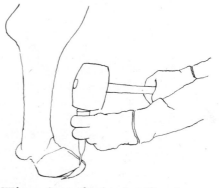

When using a chisel and mallet, always work against a wooden surface.

Presto! She's down. A large cow succumbs quickly, her legs giving out as pressure is applied to slip-knot and half-hitches around her heart girth and waist. The noose takes in one front leg as well as the cow's neck to make sure she doesn't get throttled. The cow will remain down as long as tension is kept on the casting rope. Another method for keeping a cow down is to have someone sit on her head.

Ropes and Ties

It is possible, as a last resort, to put a cow down using a stout rope that is at least 30 feet long. First place a large, slip-knotted loop around her neck *and one foreleg,* tied so that the knot is on top or just to one side of her shoulders. Run the rope back, taking a half-hitch at her heart girth and a second half-hitch in front of her udder (see photographs). One person stands holding the cow's head by a halter lead (or it can be tied to some fixed object two or three feet from the ground) and one or two people pull the long throwing rope from behind. Apparently the positioning of the half hitches and generally the application of all the ensuing pressures causes a temporary paralysis and the cow settles to the ground. As soon as the tension is released she can get up.

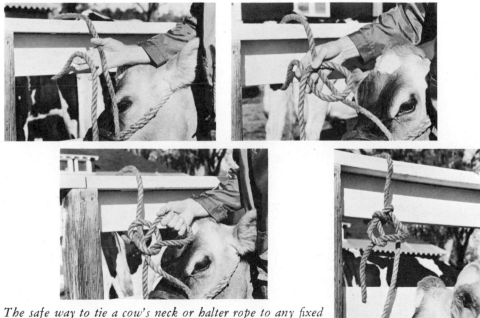

The safe way to tie a cow's neck or halter rope to any fixed object is with a quick-release knot such as the one shown here. Steps one through three are self-explanatory. The fourth is optional: dropping the loose end of the rope through the loop. This prevents the cow from working herself free, which she could do were the end of the rope left dangling where she could get a hold for a good tug.

If you wish you can tie the cow down with slip-knots of durable rope around each pastern. But be very, very careful. Those legs and hooves are potential killers.

Don't ever walk away leaving a cow tied short by a halter or neck rope unless the tie is but inches from the ground. Otherwise she might slip, fall and throttle herself. Even when you're working with a cow but have to have her tied, use a simple, single-bow knot that can be instantly untied if the cow happens to fall.

CHAPTER 6

Buildings, Staking, Fencing

The roughest kind of old barn can be cleaned out and patched with tarpaper to keep a cow. I know because I've done it, and then kicked myself for months afterward for not having fixed it right before moving the animals in.

Stalls

The floor under the cow doesn't have to be concrete, but this is best because it is tight and easy to clean. Stall and walkway surfaces should be floated off with a wooden float so that they don't become ice-slick. The next best floor is of heavy wooden planks or of creosoted blocks set on end in hot asphalt. Wood makes a warmer floor for the cow to lie on but it is more difficult to keep clean. A dirt floor in the cow barn will do in a pinch but it is impossible to keep clean, and there are heavy losses in manure value in the urine that seeps away.

Cows like a 10-by-10-foot box stall with poles or boarding for sides four to five feet high. But keeping them this way takes at least three times the bedding needed for a conventional tie or stanchion stall. If a box stall is used it is likely that you will want a conventional stall on the side for milking.

Two things the conventional stall should be: gently sloping to the rear, and long enough (but not too long) for the cow to be able to defecate in a gutter or "drop" behind.

If a barn comes with a stall or stalls that are too long, move the stanchion or tie rail back or install a *trainer.* This is any device that keeps the cow from humping forward when she urinates. It can be a store-bought electric rig, or a rigid pole fastened across the stall just above and ahead of the cow's shoulders. This system worked quite well for us.

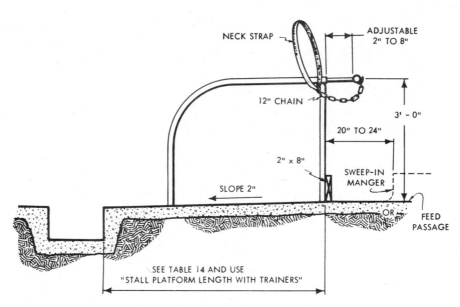

Construction details for a tie-stall.

DIMENSIONS FOR STANCHION TIE-STALLS
FOR DAIRY CATTLE

Animal Weight (lbs.)	Stall Platform Width	Stall Platform Length Without Trainers	With Trainers
800	3 ft. 4 in.	4 ft. 6 in.	4 ft. 10 in.
1,000	3 ft. 8 in.	4 ft. 8 in.	5 ft. 0 in.
1,200	4 ft. 0 in.	5 ft. 0 in.	5 ft. 4 in.
1,400	4 ft. 4 in.	5 ft. 4 in.	5 ft. 8 in.
1,600	4 ft. 8 in.	5 ft. 8 in.	6 ft. 0 in.

(From *Dairy Husbandry in Canada,* Publication 1439, 1971, Canada Department of Agriculture, Toronto, Canada.)

Stanchions

A stanchion comes in many forms, though not usually from a store any more, since the cheaper and less restraining tie systems are more popular. Often a stanchion can be picked up for little or nothing out of an old barn. They are rigid top and bottom or fastened by short lengths of chain.

For cleaning and ease of movement it's good to have five or six feet of concrete or other tight flooring behind the gutter. In a large barn this space behind the cows is called the *alley*.

A versatile barn could be set up with movable, five-foot board and post sections or hurdles, making it possible to convert a box stall into two conventional tie stalls—the one for the calving and the other for keeping the two animals.

There is no end to the variation that can go between a tight roof and floor, and barn building can be fun for people like myself who will never be fancy carpenters. Build ruggedly, arrange the space for easy feeding and cleaning. Put in wide doors and gates. Avoid leaving nails or boards sticking out where people or cows are moving around. And finish off with a good coat of whitewash.

RECIPE FOR WHITEWASH

Dissolve three pounds of common salt in 1½ gallons of water. Add 10 pounds of hydrated lime. Mix thoroughly to a thick paste. Add more water to reach the desired consistency. Spread with a coarse-bristled brush. Up to a pound of casein glue may be added to make the whitewash less flaky.

Water System

The greatest time-saver in any barn will be automatic waterers. I've never had them in my own barn, and there is really no excuse not to have done it long ago. A ditch will have to be dug from the house to the barn. A long ditch, three feet down to get below the frost line. But, as a friend said last year, "If you'd take a shovel of dirt for every time you've walked a bucket of water out there to the barn you'd have that ditch dug by now." So true. Maybe this year.

All things considered—or mostly all—a well-equipped barn for a cow and calf might have anywhere from 200 to 1000 square feet of solid ground floor space. Remember in building a barn, or in fencing a pasture, it will take less material to get around a square space than around one that is long and narrow.

Feed Storage

Tables in the appendix will help in estimating how much more space will be needed in a barn for storing hay. If shavings, sawdust or peat moss are going to be used for bedding, it helps to have a large bin somewhere near the stall. Both hay and bedding could be stored in an upstairs loft or outside under canvas.

Even under the best home-grown conditions, cows and calves sometimes get loose in the barn. For this reason always try to keep stall areas completely separated from where the hay and concentrates are stored. Don't keep grains or concentrates in bags on a floor, because this is an invitation to rats, squirrels and mice. A covered bin or large garbage cans should be used for storing these feeds.

The stall area should have two or three good-sized windows, for light and ventilation. It's bad for animals to be closed up in their own moist air and without any light. If the windows can be on the south or sunny side of the barn, fine. But don't put a window directly over or beside a cow's tie or stanchion stall. She's likely to break it, especially if she has horns, and besides, even a window with a storm sash will be the source of a cold draft.

Tethering

For grazing, cows can be fenced or tethered, or both. Often the combination works best for getting the most benefit out of a small piece of property. The cow can be tied to work along a roadside or any other small patch of ground that otherwise would go to waste. Cows make excellent lawnmowers, but their exhaust systems will ruin a good game of croquet.

Sometimes a cow gets tangled up on a tethering rope. Usually it only happens once. Then she has learned about tethers, and there's no more trouble. But if the cow is a persistent bumbler it might help to try this system: Attach a two-foot-long piece of rope from the cow's neck collar to the end of a slender spruce or other tree pole that is about eight feet long. Put a swivel on the far end of the pole and tie about 10 feet of rope from this to the stake.

A cow on a short tether, no more than 20 feet long, will make better use of available feed. This is because no matter how long the rope is, she will spend most of her time grazing the outside of the circle and much of the feed in the center will be trampled and soiled.

The tether can be any quarter-inch synthetic rope or half-inch Manila. A crowbar or an old axle makes an excellent stake that can be easily moved. Trees and posts don't make good stakes, especially for calves, because the animals tend to wrap the tether round and round until they have no room to move.

Cows will suffer and burn under a hot sun. Don't stake or pasture them where there isn't any shade.

Fencing

One to two-and-a-half acres of good soil will supply pasturage for a cow in most areas of these countries. In some cases it will take irrigation to do it. The pasture can be fenced with old cars or fancy painted boards. It can be one strand of light, common wire electrified and fastened to stakes 15 feet apart. Or it can be three or more strands of barbed wire fastened to posts on something less than 10-foot centers.

The choice depends on money, materials at hand, terrain and what kind of pasture a person has in mind. If all of the property on a small place is equally tillable there may be no desire for a permanent pasture. The one-strand electric fence on metal stakes would be easier to move.

The roughest of cleared land usually becomes permanent pasture, though in many cases it might better be let go back to woods. Fence out old garbage and junk piles, and have the area that is close to the barn or near pasture feeding bunks well drained. If you can't fence out a

A simple pole fence. It costs little but your time if you leave the trees.

junk yard, put a magnet in the cow as you would a pill. (See Chapter 17.) Sounds funny, but a magnet—and they're available through most livestock supply houses—settles in the cow's reticulum (Chapter 8) and keeps most of the iron junk a cow eats right there where it's less likely to do the harm it might were it allowed to travel along through the gut.

The most secure fencing of boards, poles or four or more strands of barbed wire will be needed between the cow and a garden or other attractions. A couple of strands may be all that's needed to keep her out of a bushy woodlot; and a steeply-banked deep river or lake may be all the fence that's needed on that side of a pasture.

FENCE POSTS

If wooden fence posts are used select one of the species listed in the appendix or take one that is locally popular. Most wooden posts will resist rot better if they are peeled, a job that is easier done when the sap is in the tree.

Green posts may be set in the ground immediately. However, they will last up to twice as long if they are loosely stacked and air-dried for three or four warm months and then treated with a wood preservative. Up-end the posts for two days or more in a barrel a quarter or so filled with creosote or a solution made up of 5 percent pentachlorophenol and 95 percent kerosene or stove oil. Throw in any old motor oil that's kicking around. Don't smoke!

Eight-foot posts four inches or more in diameter should be buried or driven about three-and-a-half feet down. Corner and gate posts may want to be longer, stouter and set deeper. A general rule in putting down wooden fence posts is that they needn't go deeper than 10 times their diameter, because in average dry soils they will snap off before uprooting from that depth.

Whether posts are driven or dropped in dug holes depends on soils and the diameter of posts. Buried posts go in the ground butt or fat-end down, and the soil has to be well tamped with a bar as the hole is filled. Posts are more easily driven in spring when the ground is moist. Drive a guide hole first with a crowbar. The narrower end of a driven post is first sharpened with an axe. It wants to be a pencil kind of sharp, neither too tapering or too blunt. It helps to have two people to do the sharpening, one to cut while the other holds and turns the end on a chopping block.

MALLETS, MAULS, SLEDGES

Use a driving maul or a sledge hammer to drive the posts. A good, homemade maul can be made out of a short hardwood log (apple is perfect) drilled through and with a length of one-inch iron pipe pounded in for a handle. The end of the handle going into the log can be split down a couple of inches with a hacksaw for wedging. Another homemade implement for sinking posts is a pile-driver made out of six-inch diameter iron pipe (see illustration). There is no fear of missing the top of the post with this tool. Extra iron weight can be welded to the top of the driver if it's wanted.

When using a maul or sledge hammer the back of a half-ton truck makes an ideal platform for the person driving the posts.

Iron post driver with 1-inch iron pipe handles.

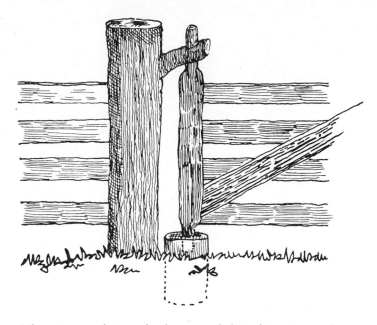

Theres' no costly iron hardware needed in this gate, a style
that is popular in parts of Europe and South America.
(Maybe in North America, too, at some time in the past.)
The bottom of the "pin" post rides in a deep socket bored
or chiseled in the end of a stout log buried in the ground.
Creosote or old motor oil can be poured in the socket to
prevent decay.

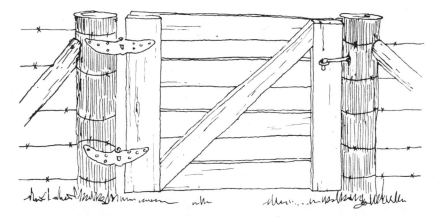

The diagonal bracing on a gate such as the simple board
gate illustrated here must run from the bottom at the hinged
side. Often homemade gates will be hung with the bracing
running the opposite direction; a sag soon follows.

WIRE

If live trees are used as fence posts, spike inch boards up their trunks and fasten the wire to these. It keeps the tree from growing around the wire. The wire will be easier to remove and won't be as likely to end up years later in a grandchild's chainsaw.

Wire alone will carry two or three posts that can't be well anchored because of mud or bedrock. If the problem is a rock outcropping on the side of a steep hill, the posts can be kept from flopping on the ground by bracing or by hanging stones on the downhill sides of the posts.

Barbed wire can be miserable stuff to handle. It's got an ornery will of its own. Leather gloves will save time and skin. Also there is a fencing tool that is to fence work what Vise-Grips are to car mechanics. This one tool, that looks a bit like a long pair of pliers with a rhinoceros snout, pulls and pounds staples and cuts, twists and grabs wire for stretching.

There aren't any bargains in barbed wire. I found a "deal" on a

When rock ledges make it impossible to sink a post on a sidehill, heavy stones hung on the downhill sides of one or more posts may keep the fence from flopping against the hill, which it wants to do because of the tension from the wire.

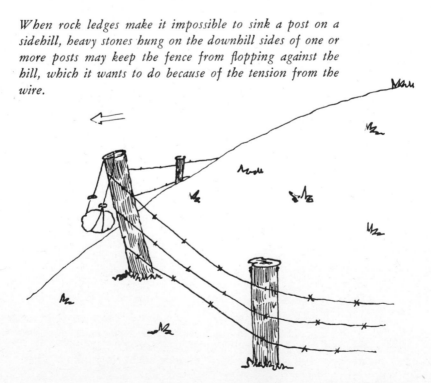

couple of rolls the other year, but one hot day a heifer ran into the fence and the wire stretched and sagged like Mozzarella cheese. And on a cold day one of the dogs snapped through a strand without breaking stride—or her hide.

Better than buying cheap new wire may be the picking up of an old fence. Unstaple the wire and fasten one end to a bumper or tractor hitch and drag it to where it can be rolled with ease. In coiling used wire, make a large loop and roll this along and back and forth across the wire stretched out on the ground, picking up from the bottom as you go. The barbs will snag and help to hold the coil together.

Two people can feed out a new reel of barbed wire very easily by walking away with a length of pipe or a crowbar run through the spool.

FENCE STRETCHING

Long, straight stretches of wire may be pulled taut with a truck or tractor, but there may be a tendency to overdo the tension, especially on a hot day. When the temperature falls the wire will tighten still more.

There are fancy little fence stretchers on the market, but the standby stretcher for small jobs is a claw hammer, hooked in the wire by a barb and hauled around the post. This, too, is usually a two-person job—one to stretch and hold the hammer while the other drives the staples.

A better homemade stretcher can be made by sawing a deep, narrow notch in the end of a hardwood board. It's better because it won't cut into the wire the way a hammer claw will, because it can be long enough for excellent leverage, and because stretching becomes a one-person job. After the wire has been pulled around the anchoring post the board can be braced behind your body while you drive the staple.

Ideally the wire on a small pasture will be stretched and tied fast only at the corners. These corner or gate posts get a lot of strain back along the line. They should be supported by posts or boards angled from the top of the end post to the base of the first line post.

ANCHORING & STAPLING

All three (or however many) wires should be anchored at a corner or gate post and stretched to the next anchoring point. Then

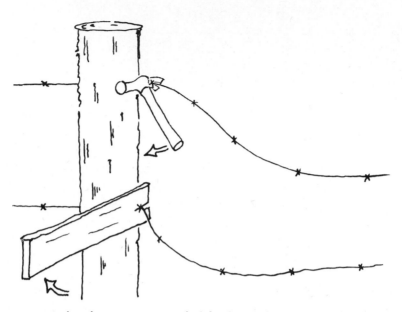

*A claw hammer or a notched hardwood board may be used
to stretch barbed wire.*

back-track, using staples or bent nails to tie the wires to the line posts
about a hammer's length apart, beginning 12 to 18 inches from the
ground. Four strands starting a foot from the ground and closer together are better than three, because the cow is less likely to "work"
on the fence as she tries to feed under, over and through. Five or six
strands are better still, if a person can afford them.

Staples should be driven on an angle to the post's grain to avoid
splitting. On-line posts' staples should be driven to the wire but not
crunched in, because the wire has to have freedom to adjust to changing
temperatures. Also, the more a galvanized wire is beaten and bruised
the faster it will rust away.

Normally, and on a straight run, barbed wire goes on the pasture
side of the posts. It should be switched to the outsides of the posts on
outside curves.

Climbing over a wire fence weakens it badly. Wherever people
frequently cross, it makes sense to sink two posts close together, nailing
board cleats across the two to form a stepladder.

Electric Fencing

An electric fence is easily and cheaply built with scrawny posts and any kind of light, common wire. The one wire (though it could be two) is usually placed about three and a half feet from the ground and is wrapped around or fastened to porcelain or plastic insulators nailed to the posts. The head of a sledge hammer held lightly against the opposite side of a flimsy post makes nailing a lot easier.

Electric fences are cheaper to put up, but if they are only made with one or two strands of common wire they are not as reliable as a good barbed-wire or board fence because there is always the chance that something may go wrong with the electrical system. Sometimes a single strand of electrified wire is used in conjunction with two or more strands of barbed wire. The electrified strand need not be stretched tight.

A single strand of electric wire for full-grown cows should be about three and a half feet from the ground. If there are calves in the pasture there will be the need for another strand about 18 inches from the ground.

The current for an electric fence can be from the house, run through a transformer and regulator that sends a non-lethal pulse over the wire. Or it can be operated by a six-volt dry or wet-cell battery. The advantage of the wet-cell is that the batteries can be recharged. A small wind generator might be just the thing for keeping one battery fully charged.

Some electric fence units send a very short, high-intensity pulse over the wire which has an advantage in that it is supposed to burn off growing grass that otherwise can ground out these fences. With other units it's usually necessary to walk the line every once in a while with a sickle or scythe to cut down the grass and weeds that may fall against the wire with the next rain.

CHAPTER 7

The Plants Cows Eat

When you look at a cow you're looking at two digestive systems with very different abilities to take plant material apart for rebuilding into bone, muscle and milk.

The first is a lot like our own, with a mouth, teeth and tube to a stomach where chemicals are poured on foods as they come along, then on to intestines where more chemicals are added and where broken-down foods are taken away into the bloodstream.

The second is the business of ruminant digestion and involves the work of billions of bacteria and other microscopic organisms that 'live in great chambers between the cow's gullet and true stomach. These creatures live on grass and other plant foods taken in by the cow, and as they live they give off heat and chemicals the cow can use. Best of all, they don't live forever. Eventually, by the billions a day, they pass away and into the depths of the cow's bowel. They are digested into pounds of nutrients the cow might not be able to get from tough and fibrous foods if she only had the first and simple, our-type of system.

However it happens, digestion takes work. Work takes energy, originating in the sun and captured by green plants through that process called photosynthesis.

These plants—the plants cows and their gut microorganisms live on—grow from carbon dioxide, water, and minerals. Most of the carbon dioxide is taken out of the air through little holes called *stoma* in the under sides of the leaves. Most of the water and minerals are taken in through the roots.

When we think of plants as animal feeds, we can divide their parts into macro- and micronutrients. The *macro*nutrients include wa-

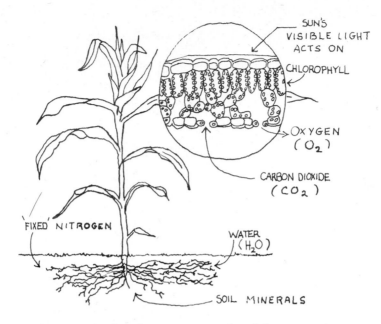

The cow's food and energy are gathered by green plants out of the sun, soil and atmosphere. Here the leaf of a corn plant can be seen absorbing visible light from the sun. The light energy excites the green chemical called chlorophyll, *which in turn enables carbon dioxide and water to combine to form the simple sugar called* glucose. *This is an extremely simplified version of what actually takes place as photosynthesis (light-building) proceeds. Hundreds of intermediate steps are involved, many of which are not yet fully understood. Nitrogen and soil minerals are needed for the synthesis of more complex organic compounds that make up plant and (eventually) animal bodies.*

ter, carbohydrates, fats and amino acids—the building blocks of proteins. The *micro*nutrients include minerals and vitamins. (Sometimes, in laboratory analysis, they are divided another way into water, organic compounds—which would include the carbohydrates, fats, amino acids and the vitamins—and minerals, or ash, since what is left over after a plant is burned is its mineral matter.)

Organic is the name for any chemical compound whose molecules are built around skeletons of carbon atoms. The carbon atoms of an organic molecule may be linked together in straight chains, branched chains, in rings or in any combinations of these shapes.

Carbohydrates

By weight or bulk the main products of plant synthesis are the carbohydrates. The most important of these in cow nutrition are the sugars, the starches and cellulose, the last being the main ingredient of tough plant cell walls.

Carbohydrates are chemically related by having two hydrogen atoms for each one of oxygen in their molecules. It's the same ratio of hydrogen to oxygen as is found in water, and so the name that means "water carbons."

SUGARS

The simplest, most easily digested carbohydrates are the sugars, and the simplest sugars come in two families that have molecules of either five or six carbon atoms. *Glucose* and *fructose* (fruit sugar, the main sweet ingredient in honey) are six-carbon sugars. Sometimes they are found tied together in more complex, compound sugars. *Maltose,* the sugar of germinating seeds, is a compound of two glucose molecules. *Sucrose,* the principal sweet of sugar cane, sugar beet roots and maple syrup, is made up of glucose-fructose pairs. The sugars are high-energy foods. They are usually found concentrated in plant juices. Sometimes they are stored in roots or fruits.

Sugar beets are rich in sucrose, a high-energy carbohydrate.

STARCHES

More often plants store reserves of energy by combining many
sugar molecules together to form starches, the next order of carbohydrates. The starch molecules are then packed away in microscopic
grains in fruits, seeds and root tubers. As nutrients the starches, too,
are considered high in energy and highly digestible. However, they
don't rate quite as high as the sugars, since in animal digestion more
energy has to be spent breaking them down into sugars before they
can be used.

CELLULOSE

Next in the carbohydrate line, but a huge jump in complexity
from the starches, comes *cellulose* and its close relatives called *hemi-
celluloses, pentosans* and *lignin*. Together they make up plant cell
walls. One molecule of cellulose may have hundreds of carbon atoms
strung together in incredible knots that only bacteria and a few other
microorganisms have the digestive enzymes to untie. As for lignin,
even a cow with her microbial helpers might as well look for nourishment in a sheet of plastic.

Any plant matter that is high in cellulose or related tough carbohydrates is called a *roughage*. It is said to be high in fiber. The percentages of these carbohydrates increases as a plant matures and dries.
Seed hulls and dry plant stems are the highest in fibers.

As a group, the roughages are low-energy foods. In cow feeding
there is a temptation to avoid them or at least to cut way back on their
use. But cows need fiber for the proper running of their digestive systems and for fatty acids that are released as gut microbes digest them.

Fats

Plants don't lay on rolls of soft fat, but fat and fat-related compounds are found throughout their bodies serving in dozens of ways.
They are called oils if they are liquid at room temperature. *Chlorophyll*
is a member of the fat group. So are the oils that give plants and

flowers their particular smells. Several vitamins are related to the fats.

Chemically, fats differ from the carbohydrates by having less oxygen in relation to their numbers of carbon and hydrogen atoms. As cow feed they have more than twice the energy value of carbohydrates or proteins (2.25 times). But in feeding cows we can't simply replace carbohydrates with fats—even if it seemed the cheaper way to go— because the cow's digestive system can only handle so much.

Although fats may be found anywhere in a plant, they are mostly concentrated in the germ and in the outer layers of seeds. The oils of some seeds—flax, cotton, peanuts, corn and others—are extracted for direct use in industry and people feeding. Leftover high-protein cakes or meals are used in livestock feeding. In some of the meals there is an added bonus in oils that escaped the extraction process.

Amino Acids and Proteins

About two dozen different amino acids are known to be produced by different plants. They are complex molecules. At a glance they might look like carbohydrates or fats but they always have about 16 percent

These Jersey cows on pasture get plenty of vitamin D, the "sunshine" vitamin that cows need for utilizing calcium and phosphorus.

of nitrogen in the form of NH_2 (amino groups) attached here and there to their carbon atoms. Many amino acids also have atoms of sulfur. Some have phosphorus and other minerals as well.

Like letters of an alphabet, amino acids can combine to form thousands of different proteins that take part in the structure and functioning of plant and animal systems.

Newly divided cells are heavy in their proportions of proteins that carry on cell growth and reproduction. For this reason the actively growing and dividing tissues in the leaf and stalk tips of young plants are excellent sources of proteins. Proteins also are concentrated in seed germs, the tiny life centers waiting to sprout when the right conditions come along. (Starch and oil deposits surrounding the germ and making up the bulk of most seeds are there to supply energy to the new sprout until it develops its own roots and leaves.) Among the forage crops, legumes have up to twice the protein value found in the grasses.

Animal and plant proteins are made out of the same amino acids, but they are put together in different ways. Also, the average animal body is higher in protein than that of any plant. One reason is that animal cell walls are made out of these chemicals instead of being made out of those tough and rigid carbohydrates.

"COMPLETE" PROTEINS AND "ESSENTIAL" AMINO ACIDS

The best proteins are called "complete proteins" because they supply all or most of the amino acids in the amounts and proportions needed for building animal tissues. Unfortunately for many animals that live primarily on plants, few plants have complete proteins. Those of soy beans and peanuts come very close. With many animals it's been found that certain amino acids are more important to health and development than others. They're called "essential amino acids" and always have to be included in the diet.

Because some "essential amino acids" are very hard to come by in the plant world, many animals—including pigs, dogs and humans— usually have to eat some complete proteins of animal origin such as meat, blood or milk. This is not the case with cows, goats or sheep. Being ruminants, they can survive handily without animal proteins—

or, in fact, without any direct source of complete proteins or "essential amino acids"—because bacteria and other simple forms of life existing in the foreparts of their digestive systems are able to build all of these nutrients out of anything that provides them with nitrogen, carbon, hydrogen and oxygen. They can do it with a quantity of the most incomplete plant proteins. They can do it with simple nitrogen compounds like urea, as long as these are mixed in the host's feed with the right proportions of easily digested carbohydrates.

Nitrogen

Nitrogen is the key to building the simplest amino acid. Our world is full of it—78 percent of the air we breathe for a start—but it is mostly in a pure state of paired nitrogen atoms that don't want to react with any other chemicals. Getting this free nitrogen into the food chain takes forcing the atoms apart so that they will recombine—become "fixed"—with other elements to form chemically active units.

SOURCES OF FIXED NITROGEN

Lightning fixes a little atmospheric nitrogen—about five pounds an acre a year—by forcing it apart and into combination with oxygen. These new units combine with moisture in the air to form acids that are brought to the ground with the next rain or snow.

Then there are bacteria and other microorganisms living in most soils that are able, under the best conditions, to fix up to 25 pounds of nitrogen per acre per year. They get their nitrogen out of the air circulating in the topsoil.

By far the most important source of fixed nitrogen comes by way of bacteria that live in the roots of clovers, alfalfa, the vetches, peas, peanuts, soy and other beans, and many other plants that belong to the legume family. They, like their free-living cousins mentioned above, take nitrogen out of the soil air and combine it with hydrogen to form ammonia, which is converted into nitrates used by the microorganisms to build amino acids. At various stages along the line, in exchange for providing a home for the bacteria, the host plants take what fixed nitrogen they need for their own amino acid and protein-building schemes.

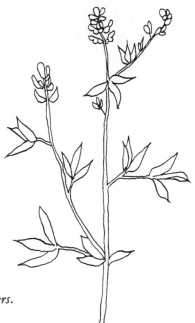

*Alfalfa is an important member of
the legume family, the nitrogen-fixers.*

Other plants too, get the benefit of nitrogen-fixing legume bacteria. From time to time the root nodules where they live break off, and grasses or whatever other plants may be growing alongside are able to pick up the bounty. When a heavy, pure stand of legumes is plowed into the ground, up to 400 lbs. per acre of fixed nitrogen may be tucked away for the next crop. Once nitrogen has been fixed into living communities, much of it remains for years and generations in a cycle of life, death, decay, and rebirth.

Vitamins

The vitamins, their chemical precursors (or provitamins), last on the list of organic compounds produced by plants, were the most recently discovered and understood—or at least partially understood. Even now, 60 years after the isolation of the first vitamin, B1, the field is littered with shifting theories and late-breaking revisions. The main problem is that the minute quantities involved tax the limits of technology, just as (at the other extreme) there are few statements to be made about the nature and origin of the universe that will stand unshaken under the eye of the next order of super telescopes.

A vitamin is any of a group of organic chemicals found in very

83

VITAMINS IN THE COW'S DIET

Vitamin	Sources	Uses	Symptoms of Deficiency	Supplements, Notes
A	Carotene and some other yellow pigments in green plants, yellow squashes and roots. In most cases the greenness of hay is indicative of good carotene content. But hay kept cool and dark may hold greenness beyond the Vitamin A value. Among the grains only *yellow* corn carries significant A value.	Necessary for the proper function of tissues producing mucous and other related fluids in the air passages, digestive tract and eyes (tears).	Night blindness is a definite indicator of a serious deficiency. Often low Vitamin A will result in frequent head and/or chest colds with runny eyes and nose, possibly leading to pneumonia. Calves may go totally blind as a result of swollen tissues pressing on optic nerve. Calves also may have severe scours. Cows will produce low Vitamin A milk, may abort or, if they go to term, may drop a dead, weak or blind calf.	Both Vit. A and Vit. D deficiencies are most likely to occur in winter, though a Vit. A deficiency can come after any long period that the cow has been largely on a diet of brown, weathered forages (pastures or hay). Quality legume hay, sun cured, would be an ideal natural source for these vitamins or their precursors. A and D *concentrates* from fish oils are widely used as supplements for these vitamins. Don't use the fish oils themselves (from which A and D concentrates are extracted) because these have various unwanted side effects, especially on mature cows in milk. Among other things they can actually depress butterfat production. Get a calf on good hay and outside as much as possible as soon as possible.
D	Direct (not through glass) sunlight acting on oily substances in plants and on animal skins converts these to Vit. D. Therefore animals on pastures or fed sun-cured forages get plenty of this vitamin. Grains and roots are lacking in Vitamin D. Green chopped forages fed fresh or as silage are not as high as when cured or weathered.	Vitamin D is necessary for the cow's utilization of calcium and phosphorus.	Rickets in young, growing animals. Swelling around pasterns (just above hooves) may be one of the earlier signs. Also see minerals table for other symptoms of calcium and/or a phosphorus deficiency.	
B (group)	With the exception of Vitamin B 12, which may be in short supply if the cow's diet is deficient in cobalt, there is not likely to be any problems with these vitamins because they are either adequately supplied in foods, are produced by rumen microorganisms, or are synthesized by the cow from nutrients found in any half-way reasonable diet.			
E				
K				
C				

small quantities in the body and necessary for the regulation of life processes. The name "vitamin" came from earlier theories that had scientists looking for vital *amines,* chemicals that are related to the proteins. When it was discovered that fats, alcohols and other chemicals were involved, they dropped the *e,* leaving a word that is ripe for exploitation by every quack doctor and patent medicine factory.

VITAMINS A AND D

Of all the vitamins and vitamin precursors so far discovered, only A and D are of major concern in cow nutrition. The rest, including E, K, C, and the B vitamins, are either abundant in any balanced and not-too-refined diet or are produced by cows or their gut bacteria from chemicals in ordinary feeds meeting the cow's other nutrient requirements.

The usual source for vitamin A is through feeds that contain its precursors, the carotenoids and related yellow compounds. These are found in yellow corn, carrots, yellow squashes and sweet potatoes. They are also found in fresh green forages and in good-quality hay (where their color is hidden by the darker chlorophyll). Bleached and weathered pastures or hay and most of the grains are deficient in vitamin A.

Vitamin D is called the "sunshine" vitamin because it is produced in animals or plants that are exposed to ultraviolet light. Cows on pasture can get vitamin D as it is produced by sunlight acting on skin oils. Some of the vitamin D is absorbed directly through the skin. Some is licked off the hair and swallowed as cows groom themselves or their friends.

But if cows are shut up in a barn for a couple of months or more where the only sunlight they get comes through glass which filters out UV rays, they begin to need some other source of vitamin D. This is especially true for growing cattle or cows in milk, since vitamin D is needed for the metabolism of calcium (Ca) and phosphorus (P). They can get this vitamin through sun-cured hay or through commercial vitamin supplements.

OTHER VITAMINS

Vitamin E comes through green feeds, quality hay and through seed germ oils. Cows produce their own vitamin C in their tissues, doing such a good job of it that they apparently don't need any at all from outside sources. All of the B vitamins and vitamin K are produced by bacteria in the cow's rumen.

Minerals

A cow needs some of more than a dozen mineral elements to keep going and to produce milk and calves. Some minerals, like calcium and phosphorus, she needs by the pound every year, for the growth of bones and the production of milk. Others, the "trace" minerals, may only be needed by the pinch, but they are just as vital. For instance, a pinch of cobalt must be in a cow's feed for the rumen bacteria that produce her vitamin B12. Other minerals cows get from plants are chlorine (Cl), copper (Cu), iodine (I), iron (Fe), magnesium (Mg), manganese (Mn), potassium (K), sulfur (S), sodium (Na), selenium (Se) and zinc (Zn).

NATURAL SOURCES

The natural way for a cow to get these minerals is by way of plants that absorb them out of the soil through their roots. For this natural way to work, the minerals have to be in the soil, both in a soluble form and in the amounts required. Crop foods have to be ones that either need the minerals for their own growth or that will pick up extras in the course of drinking their own mineral salts out of the ground.

This is a hard combination to find. After all, minerals aren't spread evenly over the earth. Even if they were, it would take a wide variety of plants to bring them all up to the cow's level. Agricultural practices that restrict animals to a few feeds grown on a few acres of ground make the situation that much worse.

The widest possible variety of food plants will do the best job of pumping minerals out of a soil and into a cow. But if: Some minerals aren't in the soil because the parent rocks didn't have them or because they were long since farmed away, or if some minerals are there but not in a soluble (available) form, or if some minerals are there but the plants that would bring them up can't be grown because of climate or

The Brown Swiss cow. Average weight, 1200 to 1400 pounds. Light gray to almost black, with a lighter streak down the center of the back. Muzzle light, nose and switch black. A white udder and white extending forward under the belly and as far as the navel is acceptable. The breed averages 11,000 pounds of milk a year, 4 percent butterfat. Originated in Switzerland. The Brown Swiss Cattle Breeders' Association of America was founded in 1880. Brown Swiss cows are the largest of the dairy breeds. They are fleshier and less angular than the others.

some other limiting factor: Then these minerals have to be added to the land or given directly to the cow in special salts or mineral mixes.

THE IMPORTANCE OF SALT (SODIUM CHLORIDE)

Sodium and chlorine are almost always given directly to farm animals, both because they are lacking in our soils and because our best (for yield, energy, and protein) cow crops don't take enough of these minerals into their systems.

COMMON MINERAL REQUIREMENTS FOR COWS

Mineral	Sources	Use	Deficiency Symptoms	Supplements	Notes
CALCIUM (CA)	Young, leafy forages, particularly legumes. Beet and mangel tops, kale, cabbages, and meat tankage or fish meals, especially when they're high in their percentages of bone.	Ca needed for formation of bones, teeth, milk. It is involved in many other chemical reactions in the cow's body as well.	Rickets developing in young, growing animals, recognized by enlarged, stiff joints, bowed long bones, humped back. In late stages there can be weakened, easily broken bones and paralysis in the hind legs. In older animals, milk production will be down. Depraved appetite for bones, wood, hair, whitewash. Older animals suffering from P deficiency will have depressed appetites, but perhaps depraved as well. Milk production down. Failure to come in heat. Aborted calves.	Ca and P supplements are listed in the appendix.	Vitamin D, the "sunshine" vitamin, is needed for the utilization of Ca and P. A lack of this vitamin increases the need for these minerals and may actually be the cause for this and other problems that at first appear to have resulted from mineral deficiencies.
PHOSPHORUS (P)	Adequate supplies in young forages, especially those grown on well-fertilized land. Cereal grains fair suppliers. Wheat bran a good source as are meat tankage and fish meal, again especially so if they're high in their percentages of bone.	Same as for Ca, though overall the requirements are a little lower.			
SODIUM (NA) CHLORINE (CL)	Some forages grown on alkali soils in arid regions may contain enough of these minerals to sustain livestock. But this is rare.	Na and Cl are required throughout the body There is Na in the blood, and Cl in the hydrochloric acid (HCl) of gastric juices.	A craving for common salt. Dull eyes, rough coat. A general loss of weight and in production.	Stock salt, NaCl, either "trace" mineralized or not, depending on other local conditions.	Stock salts to which other minerals have been added may be colored brown or blue or something else. This color is from dyes and is a marketing aid or gimmick, i.e.: nothing to do with the added minerals themselves.

Mineral	Needed for	Notes	Deficiency symptoms	Correction
IRON (Fe) COPPER (Cu)	The building of iron-rich hemoglobin (the oxygen-carrying chemical of the blood) requires a supply of Cu. Both minerals have other uses as well.	Except where soils are deficient in these minerals—or abnormally high in Molybdenum (see text)—there should be no problems for cows fed ample and balanced rations.	Anemia (pale gums, eye lids, general lethargy). With a Cu deficiency there may be diarrhea, weight loss, coarse and bleached coat. Swelling above pasterns. Cows fail to conceive and have trouble calving.	Check with agricultural representative or your veterinarian before deciding a deficiency of these is a problem and/or how best to correct it.
IODINE (I)	Needed by the thyroid gland in the neck, for production of thyroxine, a growth-regulating hormone.	Unless grown on I-deficient soils, most crops provide enough of this mineral. (See "Notes.")	Calves may be stillborn or be born with (or develop) goiters. This is seen as an enlarged neck, the result of the thyroid gland growing in its attempts to produce enough thyroxine.	Stock salt supplemented with iodine. Iodine-deficient soils are mostly in what is called the "Goiter belt." See map in this chapter.
COBALT (Co)	Needed by rumen bacteria for their synthesis of Vitamin B12 (Cyanocobalamin).	Generally found in sufficient amounts in balanced rations grown on fertile soils. To be sure, ask local agricultural representative.	Early symptoms may be like those for a P deficiency, including the depraved appetite. Anemia may develop. "Wasting disease" was one name given to the affliction, recently found to be caused by a Co deficiency.	Stock salt supplemented with cobalt.
POTASSIUM (K) MAGNESIUM (Mg) MANGANESE (Mn) ZINC (Z) SULFUR (S) SELENIUM (Se)	Although the cow's requirements for these minerals vary a great deal, they almost always are adequately supplied in any balanced ration. There may be local exceptions where soil conditions, perhaps aggravated by climatic conditions, result in plants having too much or too little of one or a combination of several minerals for proper cow nutrition.			

The two minerals are fed as sodium chloride (common salt), either loose or in blocks. Iodine, cobalt or other trace minerals lacking in local areas are commonly added to these stock salts when, as is the case with these two, the minerals apparently aren't needed for plant growth.

CALCIUM, PHOSPHORUS, AND POTASSIUM

The rock minerals needed in greatest quantities are calcium, phosphorus, and potassium. Without them cows fail. Without them crops also will fail to develop fully, and so they should be maintained in the soil through manuring, composting, and adding lime (calcium carbonate) and any other good source of the other two minerals.

Often calcium and phosphorus also are given directly to cows when we think they may not be getting enough through their feeds.

Minerals are not evenly distributed throughout the bodies of plants that take them up. Whether they are grazed or made into hay, young forage plants generally are better sources for minerals than when mature.

Leaves and stems need calcium for their own growth and development, and that is where this mineral will be found. Leafy forages, especially alfalfa or any of the other legumes, are high in calcium. A lack of this mineral in a soil will be reflected in lower crop yields. Yet even a small harvest of legume crops is likely to carry enough calcium to keep cows that are on diets favoring these feeds.

The leaves and stems of young forage crops will carry phosphorus in a decent amount if the soils where they were grown are rich in this mineral. Soils deficient in phosphorus still may yield a heavy crop of forages, but the leaves and stems may be critically low in this mineral. A grain crop deprived of phosphorus may fail to mature.

Grains are good sources of phosphorus. In some of the inner coatings or seed layers, the content may be very high. Wheat bran is one feed supplement often added to cow rations in part because of the phosphorus it contains.

Over-mature or harvested crops left too long in the field will lose minerals through weathering. Phosphorus, for instance, is rapidly destroyed. Hay crops that are initially low in minerals can become deficient through sun bleaching and rain. Even standing crops of late-

MINERAL DEFICIENCIES IN
THE UNITED STATES

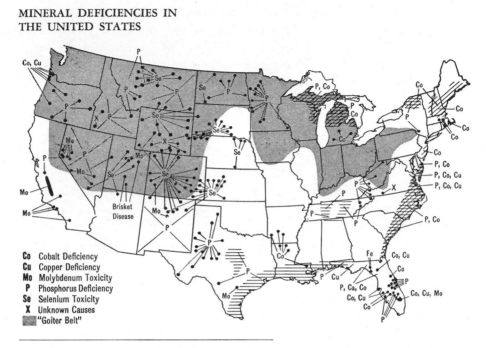

Co   Cobalt Deficiency
Cu   Copper Deficiency
Mo   Molybdenum Toxicity
P   Phosphorus Deficiency
Se   Selenium Toxicity
X   Unknown Causes
     "Goiter Belt"

Map shows areas in the United States where mineral-nutritional diseases of animals occur. Dots indicate approximate trouble areas. Lines without dots show
a generalized trouble area.

(From USDA *Yearbook of Agriculture*, 1957.)

summer pasture forages or crops like corn or sorghums for silages lose
phosphorus as they brown and die.

HOW MUCH IS NEEDED?

There can be a fine line between too little and too much of an
available mineral. Sometimes the line is the same for plants and the
cows that feed on them. The plants don't thrive and so the cows don't
thrive.

At other times plants may accept a mineral at a level above or
below what is safe or adequate for the cow and remain healthy. When
this occurs, farmers may be caught completely off guard, because there
is no indication of trouble until the cows have spent weeks or months
eating the over- or under-charged feeds. Low-phosphorus forages fit
this category.

Selenium is an even trickier example. Plants don't seem to need it, but it is a trace requirement for cows. And yet more than a trace can be toxic. Some plants readily absorb whatever selenium is available in the soil. If there is *none* available the plants are fine, but the cows or their calves can get sick. If there is lots available, the plants are fine but the cows can be poisoned.

An unusual level of one mineral may upset the normal requirements for another. In some Rocky Mountain valleys, plants pick up too much molybdenum (Mo), a trace requirement for plants but evidently not a requirement in cattle. It doesn't seem to bother the plants if they get more Mo than they need but cows eating the plants can develop a nutritional anemia—because too much molybdenum blocks the cows' uptake of copper, and without copper they can't use the iron necessary to build red blood cells.

This kind of back and forth interrelationship between one mineral and another, between minerals and vitamins, and between vitamins, minerals and all the rest of the processes that go into plant and animal metabolism is what makes the science of feeds and feeding fascinating and also impossible to deal with in pin-pointy ways.

Nutritional diseases caused by local soil mineral deficiencies used to be more common when livestock were restricted to feeds grown on the home farm. Today's small farm probably buys at least some feed supplements made from plants grown in many different parts of the continent. Too, most commercially mixed concentrate feeds are fortified with minerals.

Today we also have the help of government soil testing laboratories and the knowledge that has been built up over the years about mineral deficiencies—where they occur and how to get around them.

CHAPTER 8

Taking It All In

Humans can live on vegetables alone, and many do, though they do so most often and most successfully in warm climates where there is more apt to be a wide variety of fresh plants available throughout the year.

Cows more easily live on plants because their two-part digestive system mentioned in the last chapter is geared to these foods. It is a more complicated gut than the type needed to digest meat and other foods of animal origin. Only a digestive system with the cow's wide range of abilities can squeeze the nutrients out of tough plant cells and tissues.

The cow's use of plant foods begins with *ingestion,* a taking-in of nearly whole conglomerations of cells or tissues. Next comes *digestion,* a long and progressively refined mechanical and chemical breaking down of the tissues and cells. These are the mouth-to-anus processes of a digestive tract that is more than 100 feet long.

Then there is *absorption,* the movement of usable, fractionated foods from the digestive tract into the animal's blood stream and on to wherever it's immediately needed or can be stored for hard times.

The last step is the use or *synthesis* of the foods by the individual body cells that are growing, reproducing or actively working and releasing energy for the animal's tasks of finding more food and water, breathing and reproducing.

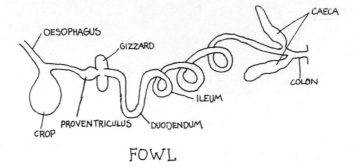

FOWL

A comparison of four types of digestive systems. Humans and dogs have systems more closely related to that found in pigs than to those found in cows, sheep (or goats and other ruminants), horses or chickens.

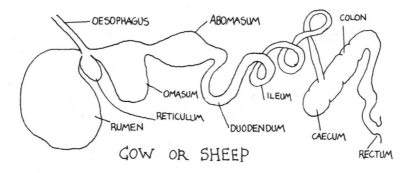

COW OR SHEEP

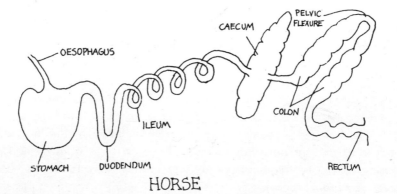

HORSE

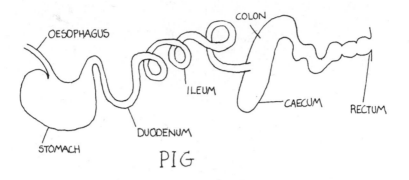

PIG

Teeth and Tongue

On the mechanical side of digestion: Cows have large, grinding molars, top and bottom, that can pulp the toughest grass or straw to the consistency of raw spinach run through a blender. We have molars too, but they aren't the type (and we haven't the jaw muscles to force them) to reduce grass to something we could swallow in less time than it would take us to starve.

Two other aids cows have (and we haven't) are their sandpaper-like tongues for directing grass traffic, and salivary glands that can pump incredible amounts of lubricating fluids—up to fifteen gallons a day—in and around each wad of plant fiber so that eventually it becomes as slippery and swallowable as canned peaches.

As a cow is feeding along through a pasture she actually uses her molars very little. Saliva and rough tongue are more important at this stage, because with them she manages to swallow great quantities of partially-chewed roughages that are packed away in a mixing caldron called the *rumen* or *paunch*. Left to herself on pasture or offered roughages free choice she will normally feed about five times a day—three times during the day and twice at night.

Sometimes the rumen is called the first stomach. And sometimes it is said that a cow has four stomachs altogether, which is a bit misleading. A cow has only one true stomach, just as we have only one true stomach. What the cow has (and we haven't) are three antechambers (rumen, reticulum and omasum) housing those billions of bacteria and other microscopic creatures which unwittingly devote themselves to the cow's well-being.

The Rumen

The first chamber, the rumen, is by far the largest. It's absolutely huge in a mature cow—able to hold 40 to 60 gallons of food, fluids and gases. It lies along the cow's left side, reaching almost the whole way from beneath the ribs back to the hind quarters.

In reality a picture of the rumen, or for that matter any part of the long digestive tract, "lying" somewhere or other also is misleading. There is constant movement, sometimes slow, at other times quite active, all along the gut. This involuntary muscular action called *peristalsis* passes from front to back, front to back, constantly mixing, turning and pushing everything but the foods needing another go-round with the molars topside, to the back of the house.

Barely-chewed forages, then, are swallowed into the rumen. This is bruised grass, busted here and there, but for the most part intact. Microorganisms are waiting, for this is their staff of life. After an hour or two of grazing the rumen tank is loaded, and much of the once-munched forage has begun to feel the softening effects of enzymes produced by the microorganisms. The cow finds a quiet spot and lies down. She's there but a moment before, *errrrp*—up comes a cud, a fist-sized slurp of green mash pushed up from the rumen and in part from the reticulum, by a neatly controlled reverse swallow. Rumination has begun. She presses excess juices and saliva from the slurp, swallows this in normal fashion, and begins to chew. Notice the side-ways grinding action. Stand by a cow and listen to the powerful, and methodical, crunch, relentless crunch.

After about a minute of this treatment the cow swallows, hangs in suspended animation for a moment, then returns with the next quiet belch and cud. Through a process that isn't fully understood, there is a sorting that goes on at the base of the throat or esophagus, that sees plant juices and the finest particles of fiber passed on to the omasum, or third chamber, while larger particles are returned to the rumen and reticulum for another dose of fermentation and the possibility of several return trips to the ivory mill.

For some reason mature cows do not chew certain foods as well as they do others. They are especially sloppy about chewing whole grains of corn, oats, barley and the like. These seeds are tough coated. But bacterial fermentation would break them down if it weren't for their small sizes and weight that lets them slip too quickly through the fermenting chambers. Up to 25 percent of whole grains fed to a cow may pass all the way through her system undigested.

Two things are done to improve this situation. One is to roll or grind the grains before feeding—although not to a fineness that makes them powdery, then pasty in the mouth and difficult to eat. The other is to make sure cows are fed plenty of roughages right before, after, or with their grains. The idea is that in this way the grains will get caught up in the roughage and more likely get the benefits of cud chewing and rumen fermentation.

The Ayshire cow. Average weight, 1200 pounds. White with red markings or predominantly red with white. The red may vary from light to almost black. The pattern is usually more broken and spotty than is seen in the red or brown and white markings on some Jerseys, Guernseys, or "red" Holsteins. Production ranges from 11,000 to 12,000 pounds a year and butterfat averages 4 percent. Ayshires originated in Scotland. The U. S. Ayshire Breeders' Association was founded in 1875.

The Reticulum

Although cows may refuse dirty hay, they aren't against swallowing stones, nails, baling twine and all sorts of other junk. These seem to collect in the reticulum, and so it's sometimes called the "hardware stomach." The reticulum is very small compared to the rumen. It may hold no more than three gallons, including hardware, at a time. Its queerly sectioned lining finds its way into the meat market under the name of *tripe*.

No nourishment reaches the cow through her reticulum, but a lot reaches her through the walls of the rumen and omasum—also known as the *maniplies* for its many deep folds. Most of what she gets through the rumen and omasum are fatty acids including acetic, butyric and propionic acids given off in the process of gut fermentation. Low-roughage diets can lower the quantities of these acids given off, or change their proportions in ways that may lower a cow's production of butter fat. This wouldn't be a problem with a family cow on unlimited pasture or any diet that included a minimum of 10 pounds daily of long (not chopped) hay or grass.

The Abomasum

The action of the rumen, reticulum and omasum microorganisms ends in the omasum. If they haven't died of old age by this time most will be killed by the high-acid gastric juices, including hydrochloric acid, secreted by the walls of the abomasum.

In addition to the HCl, the first important digestive enzyme is produced in the abomasum wall and mixed into the off-green collection of foods that by this time have the consistency of a porridge. This is the enzyme *pepsin,* that begins the breakdown of proteins into pieces called *peptides.*

The Small Intestines

Now the food, further churned and racked by the acids and enzymes of gastric juice, begins to empty into the first end of that long,

ENZYME DIGESTION IN COWS

Food	Enzyme	Secreting Gland or Organ	Steps in the Break-Down
Starch	Pancreatic Amylase	Pancreas	Starch → Maltose
Maltose	Maltase	Intestine	Maltose → Glucose
Sucrose	Sucrase	Intestine	Sucrose < Glucose / Fructose
Lactose	Lactase	Intestine	Lactose < Galactose / Glucose
Protein	Pepsin (& hydrochloric acid)	Stomach (Abomasum)	Protein → Peptides
Protein (Milk)	Rennin	Stomach (Abomasum)	Casein → Milk Clot
Protein	Trypsin	Pancreas	Protein → Peptides → Amino Acids
Fats & Lipids	Lipase	Pancreas	Fat < Fatty Acids / Glycerol
Fats & Lipids	Lipase	Intestine	same as Pancreatic Lipase

(From USDA *Yearbook of Agriculture*, 1939.)

coiled digestion and absorption tube known as the small intestines. Almost immediately the *chyme,* as the mix is now called, is doused by bile from the liver (by way of the gall bladder) and at least three different enzymes from the pancreas.

Bile is a strong base. Its main purpose seems to be to break fats and oils into tiny droplets, some of which can be absorbed as-is through the intestinal wall. Other fat droplets are broken down farther into fatty acids and glycerol through the action of the pancreatic enzyme, *lipase*.

The enzymes trypsin and amylase also are secreted by the *pancreas*, a long, flat and narrow organ that opens into the intestines across the way from the bile duct.

Trypsin's job is to provoke the breakdown of peptides into amino acids. *Amylase* attacks plant starches, turning them into compound sugars. (In humans and pigs the conversion of starches begins in the mouth with the action of ptyalin that's found in their saliva. But this enzyme isn't found in cows or horses.)

At this point in digestion the fats and proteins are ready for absorption through the gut wall. Only the compound sugars must be reduced to absorbable, simple forms. Three sugar-digesting enzymes that help here are secreted by cells in the walls of the small intestines. One of them, *sucrase*, helps turn molecules of sucrose into one molecule each of glucose and fructose. Another, *maltase*, helps convert maltose, a sugar found in germinating seeds, into molecules of glucose.

The third, *lactase*, is found most abundantly in young nursing animals and is an enzyme that helps turn molecules of lactose (milk sugar) into a molecule each of galactose and glucose.

The absorption of digested foods goes on through the surface of the small intestines, made huge by an inner lining of millions of tiny fingers called *villi*. Each one is lined with its own supply of finest blood and lymph vessels only a cell's width away from the amino acids, fatty acids, glycerol, fat droplets, vitamins, minerals and simple sugars the cow needs.

The Large Intestines

Few nutrients are absorbed through the smooth walls of the large intestines. Mostly this is where water is taken out of what is left of the chyme. It's a critical process, though. Any infection in the large bowel, from bad bacteria, viruses or larger parasites, can bring quick disaster. The intestines become swollen and inflamed. They stop absorbing water, and the cow has diarrhea. She is losing water. If the problem isn't taken care of she can wilt and die like a flower in a drought.

The blood and lymph vessels surrounding a cow's digestive system are like plant roots in a rich soil. The blood and lymph are saps that carry absorbed nutrients through the body where they are snapped up by greedy, growing and working cells.

Sources of Energy and Protein

The sugars are immediate sources of energy. They can be used to build *lactose* (milk sugar). They can go into the formation of fats, or into the production of a ready-reserve source of energy called *glycogen* (animal starch) that is stored in the liver.

The fats and some of the more complex fatty acids are secondary sources of energy. Or they can be combined into the animal fats that go into milk (butterfat). Newly formed fats can be involved in energy-exchange reactions or in the transport of different materials farther on through the various tissues of the body. Excess fats can be laid down as insulation and padding to be used another time when foods are scarce.

Amino acids are taken for the proteins of cell walls, of muscles, hair, horns and hooves. Others form the proteins of milk, of hormones, blood, lymph, bones and enzymes. Some of the vitamins and all of the proteins that program and control cell metabolism and reproduction are made from amino acids.

If a cow gets more amino acids than she needs, the extras can be broken down within the body to provide as much energy as normally comes from carbohydrates.

Although minerals will be concentrated in bones, teeth and in milk, every cell or fluid in the body needs a share of these elements. They are the nuts and bolts of structure and the conductors of energy and nerve impulses through the body.

The Excrement

The feces that collects at the end of the gut are made up of undigested or undigestible foods, of good bacteria and other microorganisms that began to flourish once again in the large bowel, and of dead cells that were sloughed from the inner walls of the digestive system in the constant business of rejuvenation. There also should be enough water left to keep the animal from being constipated, a bit of a problem sometimes when cows are on a diet that's mostly dried roughages.

Although feces and urine are usually lumped in a cart and called manure, they are very different. Except for the sloughed cells of the digestive system walls, most of what's in feces never was a part of the cow's metabolism. On the other hand, the valuable fertilizing ingredients of urine are compounds and minerals that were for the most part built right into the cow's tissues and were taking part in all of the chemical changes that keep her going.

These products of catabolism (breakdown) could be reintroduced to the large intestines so that all wastes and want-nots could be eliminated at once. It's done that way in birds and duckbilled platypuses. But instead, after being filtered from the blood by the kidneys, they go to the bladder and out of the body by way of the urethra.

Calf Development

I mentioned lactase, the enzyme of milk sugar digestion. It is found in nursing calves and is only one of many differences, both chemical and physical, between their digestive systems and those of mature cows or bulls.

The first difference (and the only one that can be seen) is that normally the calf is born with only the two middle incisors in its lower jaw. But this is a small thing. Within a few weeks the calf will have all eight of her baby incisors but still be unable to eat or use much in the way of regular cow foods. One reason is that the calf is born with a small and inactive rumen.

At the same time the calf's abomasum is relatively larger than in a mature cow and secretes yet one more enzyme not normally found in an animal's system after weaning. This is the enzyme *renin*. In combination with hydrochloric acid, renin coagulates milk so that as a solid, it will stay in the abomasum long enough for the pepsin to begin work on the milk proteins.

Calves will begin to nibble hay or grass within a week or two of birth. It's fun to watch. Never a mouthful, only a blade or two at a time is dragged around half the morning before the youngster finally manages to chew it down. Slowly the rumen develops and expands.

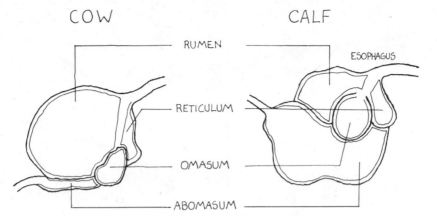

The relative sizes of abomasum (stomach) and rumen (paunch) change dramatically as a new-born calf matures. Notice how the young calf (right) with its comparatively large abomasum and small rumen is equipped for a diet of milk. It takes at least six weeks and the provision of good, long-fibered roughages for the rumen to develop to the point that the calf can carry on without milk or milk substitutes.

Given access to good roughage and some grains a calf should have a system capable of going it alone without milk or a commercial milk replacer by the age of two months.

From this time, until the age of six or eight months, calves are better grain chewers than their mothers. It isn't necessary to grind or roll their grains. Then again, neither will it hurt.

Although the calf's rumen may begin to work two or three weeks after birth, there may not be synthesis of enough B and K vitamins until at least the age of weaning. Neither will the calf be producing the adult enzymes for digesting sucrose or starch. A Georgia bulletin says some milk replacers high in cereal grain starches are poor products for this reason.

The following pages 104-7 show anatomical plates of dairy cattle. Chapter 9 begins on page 108.

Explanation of Plates on Cattle

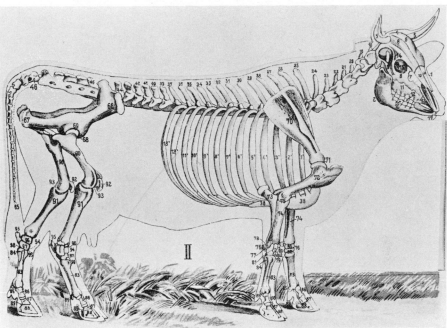

II

VEINS

74 Anterior vena cava
75 Jugular
76 Azygos
77 Posterior vena cava
78 Hepatic
79 Portal

Muscles

1 Levator labii
2 Zygomatic
3 Levator labii proprius
4 Pyramidal muscle of the nose
5 Cheek muscle
6 Depressor of the lip
7 Masseter
8, 9 Sterno-cephalic
10, 11 Orbicular of the eye
12-14 Upper, middle and lower adductor of the ear
15 Parotid
16-18 Brachio-cephalic
19 Clavicular part of same
20 Cleidomastoideus
21 Superficial pectoral
22, 23 Trapezius
24 Levator of the shoulder blade
25, 26 Infraspinatus
27, 28 Abductor of the humerus
29-31 Forearm extensors
32 Latissimus dorsi
33 Oblique abdominal
34 Large serrate
35 Deep pectoral
36 Extensor of the radius
37-39 Extensors of the foot
40 Carpal retractor
41 Oblique carpal extensor
42-46 Flexors of the foot
47, 48 Carpal extensors
49 Flexor of the pastern

50, 51 Ligaments
52 Large croup muscle
53 Tensor fasciae latae
54 Extensor of the patella
55, 56 Abductor femoris
57, 58 Adductors of the femur
59 Tail muscle
60 Flexor of the tibia
61 Toe extensor
62, 63 Peroneal muscles
64 Toe flexor
65 Gastrocnemius
66 Ligaments
67 Inner gastrocnemius
68 Posterior tibial
69 Tendon of Achilles

Internal Organs

1 Cerebrum
2 Cerebellum
3 Optic lobes
4 Medulla oblongata
5 Spinal cord
6 Cross section of the vertebral column
7 Ligamentum nuchae
8 Turbinated bones
9 Pharyngeal cavity
10 Larynx
11 Trachea
12 Thyroid gland
13 Bronchi
14 Sternum
15 Right lung
16 Pleura
17, 18 Diaphragm
19 Left ventricle
20 Right ventricle
21 Pulmonary artery
22 Aorta
23 Coronary artery
24 Left auricle
25 Right auricle
26 Apex of the heart
27 Mouth cavity
28 Tongue
29 Gums

30 Pharynx
31 Gullet
32 First stomach or paunch
33 Left sac of same
34 Right sac of same
35 Pillars of same
36 Villi of same
37 Gullet opening of same
38-40 Divisions of same
41 Spleen
42 Opening to second stomach
43 Second stomach or reticulum
44 Connection of first and second stomachs
45 Third stomach or omasum
46 Connection of same with fourth
47 Fourth stomach or abomasum
48 Pyloric end of same
49 Duodenum
50 Mesentery
51 Jejunum
52 Ileum
53 Connection of same with caecum
54 Caecum
55 Colon
56 Rectum
57 Anus
58 Liver
59 Right kidney
60 Bladder
61 Neck of bladder
62 Outlet of same
63 Vagina
64 Body cavity
65 Pelvic cavity
66 Udder
67 Glandular substance of same
68 Milk ducts
69 Milk cisterns
70 Teat tube

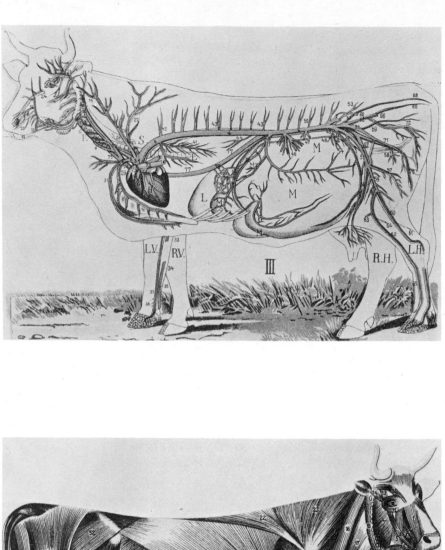

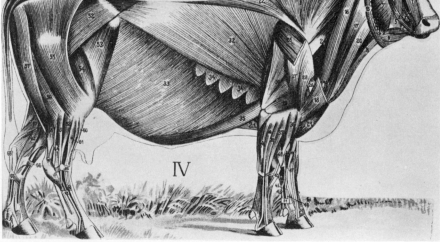

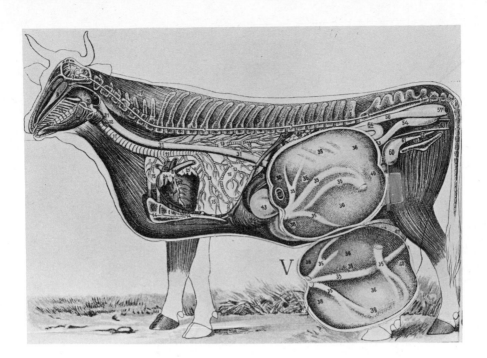

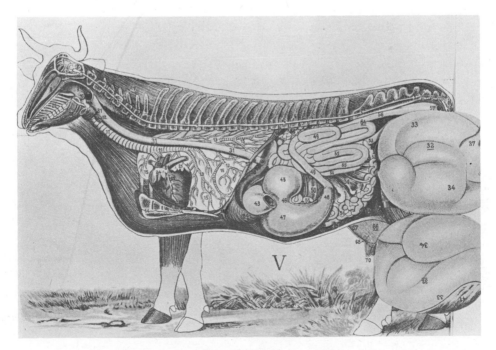

CHAPTER 9

A Choice of Foods

Everything a cow needs is in a fertile pasture of young, actively growing grasses, a block of trace-mineralized salt and a tub with a hose trickling fresh water.

The average cow may eat up to 100 pounds a day of fresh green grass. On that she can keep well and produce 20 or more pounds of average-fat milk.

She wouldn't even need much extra water since fresh grass has a moisture content of 70 percent or more. Keeping dairy cattle would be so simple . . . if the world were rolling meadows of young, actively growing grass 365 days a year.

Unfortunately this food, with its excellent proportions of digestible and palatable nutrients, is available in most of Canada and the United States only six or eight weeks out of the year on any one farm. Grass is a cool, wet weather plant. It's a spring plant with a bit of a fall rebound. This is especially true of our native grasses. Wild bison took advantage of this pattern by migrating north and south each year following the best growth.

An ideal in feeding is to come as close as possible to duplicating what young pastures have to offer, throughout the year and in one spot. The feasibility of cows is that they are tolerant beasts and will adapt to a wide variety of feeds and feed by-products that can be used to duplicate grass. They can store many nutrients during the pasture season to help them through the bad times. And they can remain at least reasonably healthy on quite a bit less than an ideal ration.

Grass Replacers

If I ran out of money in the middle of a northern winter, and couldn't buy grains or other supplements to a poor quality hay, I'd feed the cow vegetable scraps from the kitchen. I could turn her loose to browse tree buds, or bring hardwood tree "loppings" to her, as they did for years before there was the wealth and technology for growing and storing large quantities of winter feeds for cattle.

No doubt the cow would stop giving milk. It would be a return to an old way of farming for summer milk only. Spring calf, summer milk, and then off to the butcher or through the winter on a diet of roughages that little more than sustains a cow, with just enough extra for the developing calf. It's a cheap way to produce milk, letting those pastures do it for as long as they can. The season could even be extended by special plantings of spring and late summer or fall crops.

But winter milk is no longer hard to produce, and it does not have to be expensive. All it takes is a green grass replacer of good hay (or in the far South ample dry pasture and more emphasis on special crops: see Chapter 12), and a few pounds daily of concentrated carbohydrates, fats and proteins. Other foods, too, may be used to replace or supplement part of the hay or concentrates in a cow's ration.

Dozens of questions come into play in choosing which foods to use as grass replacers. These include palatability, bulk, nutrient value, possible side-effects of otherwise valuable foods, and comparative costs.

PALATABILITY

Some economical and nutritious foods lack palatability. Some simply don't taste good to cows. They either have to be left out of rations altogether or be added in quantities that can be masked by other, good-tasting foods. Buckwheat and rye are examples of grains that cows don't much like, but which can be used singly to make up to one-third of a concentrate mixture. Sometimes otherwise good foods are too coarse or too fine or dusty for a cow's liking.

Sweets or spices may be added to foods cows don't particularly like. Fennel (fenugreek) and/or molasses are often included in con-

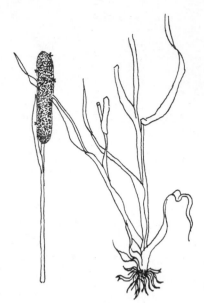

Timothy, an important hay and pasture grass throughout much of North America.

centrates to perk appetites or to hide unpopular ingredients, and salt may also be added to concentrates to increase palatability.

Molasses can be used to overcome dust problems. At the other extreme there was a farm where we used to sprinkle diluted molasses on coarse hay to encourage the cows to clean it up. It was a success. But then I tried it recently with my own cows and it was as if I'd sprinkled their hay with motor oil. The same hay they had been eating fairly well they wouldn't touch at all.

It could be that if I had kept up the molasses treatment for several days in a row they would have learned to love it. Probably so. It's in the nature of cows to balk at anything strange, even a food they may soon acquire a taste for.

Although it is no way to health and production, starvation will also get cows to eat things they don't like. They'll eat foods that have no nutritional value at all. The paunches of starved deer may be packed with pine needles they can barely digest and which under better circumstances they would only nibble in the course of browsing preferred hardwood buds and twigs.

A starved cow is more likely to eat poisonous foods. There are more than 30 species of plants across the United States and Canada that always or under special circumstances may be poisonous to cows. What a miserable situation pasturing cattle would be if it weren't that in most cases the cows will avoid the bad plants unless they are desperately hungry.

110

BULK

Hypothetically speaking, there is another way a deer or cow could starve on a full belly and that's if her diet were restricted to extremely bulky foods. Bulky foods take up a lot of room for the nutrients they contain. Bulk is an important consideration when working out rations because cows, like the rest of us, can eat only so much total quantity of stuff in a day. If the total quantity contains an unusual amount of air or water that takes up space without providing food value, milk and fat production suffer immediately. Before long the cow herself begins to fade away.

Some bulk is necessary in a cow's diet because nature gave her a digestive system that's geared to work against a mass of food materials.

In determining the right level of bulk, nutritionists work with measurements of *dry matter* (DM) in the various feeds. It has been found that cows can consume up to three-and-a-half pounds of DM per day for each 100 pounds of body weight. It would be very unusual for a cow to eat this much in roughages alone, because these are bulky feeds. Cows will eat up to 150 pounds a day of pasture grasses averaging around 20 percent DM, the amount consumed depending in part on the cow's size and the quality of forages. They will eat 15 to 30 pounds a day of dry grass or hay having a DM content of about 90 per-

Fenugreek—the spice cows like.

cent. On average then, we're talking about cows eating one-and-a-half to two-and-a-half pounds of dry matter in roughages a day per hundredweight.

Offered all the average quality dry hay she wants, it's been found that the average dairy cow under most circumstances will eat about two pounds per day, per hundredweight. This leaves plenty of room, at least in a physical sense, for 10 or more pounds of dry matter which could be filled with concentrates or other foods.

Just how much roughage a cow will eat in relation to her size depends on its quality, how it is being fed and on the demands being placed on the cow for growth, for reproduction and for the production of milk and butterfat. Genetics plays a large part, too, as it determines a cow's ability to respond to these demands.

NUTRIENT VALUES

Low-quality pastures (weedy, drought-stricken, weathered or dormant), late-harvested grass or legume hay—all can have three strikes against them as feeds. One, they will be lower in protein and minerals than the same forages harvested or grazed at earlier stages of growth. Two, they will lack the palatability of earlier-cut or grazed forages, in part because of their lower nutrient value and in part because of their coarser textures. Three, these feeds have a slow rate of digestibility because of their higher proportions of those tough carbohydrates, including cellulose and lignin.

More food value can be got out of over-mature, slowly digestible hay or straw if it is fed more frequently and in small batches. Overloading the cow's system with quantities of such extremely high fiber foods, as would happen if she was fed mounds of the stuff only two or three times a day, would be like trying to run an unwound spool of thread through a soda straw.

By laboratory analysis it can be discovered that a food contains so much of this, that or the other nutrient. But none of them will be 100 percent available to the cow. Sometimes the difference can be blamed on the physical or chemical form of the nutrient that makes it unusable. Other times it is simply in the nature of the cow. She isn't ever 100 percent efficient. She's like a wood stove with a large, open grate. Before it can extract all the heat that scientists say is contained in a stick

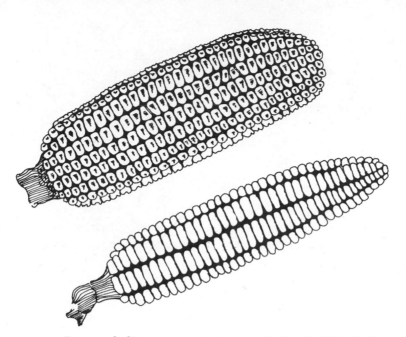

Dent and flint corn are popular varieties of field (feed) corn in North America. Mature cows may waste up to 25 percent of whole grains such as corn if they are not first ground or rolled.

of wood, the stove releases unburned gasses up the chimney and charcoal into the cool ashes below. Some fuels stay in the cow's system and are used more completely than others. Some we can improve, as when we grind or roll whole grains.

This idea of digestibility is more important on large, commercial farms where a 1 or 2 percent difference between feeds could add up to a large dollar gain or loss over a year. For the family cow owner it's useful in comparing feeds to know that there are these differences. When people talk about how much protein or fat or whatever is in a feed, find out if they are talking about the grand total or the amount that is available to the cow.

Crude protein (CP) or *protein* without any qualifiers are terms used for the total protein value in a food. *Digestible protein* (DP) refers to what is there for the cow.

Usually there is less ambiguity with the other nutrients, the prefix "digestible" being used if that's what is being considered.

Two foods having equal amounts of digestible nutrients may have different nutrient values because in the case of one it takes more time and/or effort to get the nutrients into the cow's system. The same food will have more value if it is placed in front of the cow than if the cow has to chase all over a pasture to get it. But in a case like that you have to weigh the cost of cutting and hauling the food to her.

Less obviously, one food may take more chewing than another. Or one food may have a higher proportion of those fibrous carbo-hydrates that in themselves are hard to digest and which in some cases may be surrounding otherwise easily digested proteins, fats and the like.

SIDE EFFECTS

Some feeds that are cheap, tasty and nutritious can take a dive be-cause, when they are fed in quantity, they somehow knock the cow's system out of balance. Maybe they cause diarrhea (scouring) as can hap-pen with cows fed too many potatoes. Others fed in too-large quantities can alter the quality of the butterfat so that the butter comes out brittle or greasy.

In general all of these problems can be avoided through variety grain, hay, legume or whatever, any more than you would confine in feeding; by not expecting the cow to get by on any one species of yourself to an exclusive diet of soybean cakes.

COMPARATIVE COSTS

Whether a food can be raised, bought or scrounged, "availability" is easy enough to figure. Whether the cost is right will depend on how

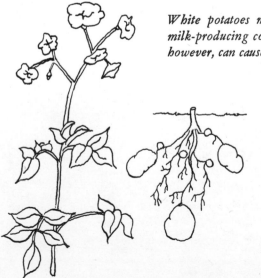

White potatoes make excellent food for milk-producing cows. Too many potatoes, however, can cause diarrhea.

each food compares to what else is available. Good grass plus a third or more legume hay could easily be worth 10 to 20 percent more than the same quality grass hay alone, because the legumes are generally richer in proteins and minerals. For this and other reasons an even larger spread in value can exist between poor-quality, overmature and/or weathered hay, and hay that comes from grasses cut early and rapidly cured.

Good commercial concentrate mixtures of grains and protein or protein sources are worth about 1.75 times as much as equal weights of good hay. Whenever the cost of good hay drops below this ratio to the cost of good concentrates—and it's been there several years now in spite of rising hay costs—it may pay to go out and buy a ton or two of excellent hay so that you can cut back on the concentrates.

Under most circumstances purchased succulents will be worth a third or less the cost of an equal weight of dried roughages because their extra weight is mostly water. For this reason it seldom pays to truck succulents any distance.

Succulents

Many farmers, especially those with large commercial operations, replace much of the hay with wet roughages or other foods of a watery nature that together are called *succulents*. The wet roughages include silages, fresh, green-chopped crops and vegetable processing wastes. The other succulents, too low in fiber to be considered roughages, include fruit processing wastes and root crops such as potatoes, beets or mangels.

Succulents are good for cows but they are not nutritionally necessary if the animals are already getting plenty of good-quality hay and a measure of concentrates that have been balanced to fill in where the hay may be lacking. Farmers like succulents for their food value, for their laxative nature and, in the case of some, for what is called their "conditioning effect"—the general look of health and vitality that comes with their inclusion in any diet.

But there's more to it than that on the large farm, particularly in humid regions where making high-quality hay for 100 or more cows can be a nightmare. The silages—for these are the most popular succulents—can be harvested rain or shine. Also, they pack and store in

The Guernsey cow. Typically weighs 1000 to 1200 pounds. Fawn to nearly red with variable white patches, especially on the legs and underbelly. The nose is described as being cream or buff colored. I see it tending to dull brick red. Certainly very different from the black nose on a Jersey or Brown Swiss. Production averages 10,000 to 11,000 pounds a year of milk with a butter level of about 4.9 percent. Guernseys are least efficient converters of yellow carotene into colorless vitamin A. The result is a distinctly yellow skin, butterfat and body fat too. Not great for producing prime veal. Guernseys originated on the island of Guernsey in the English Channel.

smaller spaces than are needed for hay. And every process from planting through feeding now can be done by machine.

Most family cow owners do, and will continue to, stick to hay as the only roughage for winter or barn-season feeding. And this is fine so long as there is plenty of fair-to-good quality hay on hand. People who are limited to low-quality hay or who are interested in experimenting with ways to cut back on the use of grains may want to grow some roots. They're sometimes called "watery concentrates" because they are so low in fiber and high in easily-digested starches and sugars. For more information on feeding succulents or roughages other than hay, see Chapter 12.

Commercial Concentrates

The concentrates used to boost the cow's daily intake of nutrients beyond what she would consume in roughages (possibly including succulents) are also called "short feeds" or "formula feeds." Good commercially prepared concentrate mixtures have guaranteed protein values and guaranteed maximums or minimums of fiber, fats and carbohydrates. Many of them supply extra minerals and vitamins as well. They may be sweetened with molasses, and sometimes the entire mixture will be steamed and formed into pellets.

No good grain or concentrate mixture should have more than 11 or 12 percent fiber. Feed stores and mills often sell bags of single or mixed ground grains carrying the names of the seeds included, but with no statements or guarantees as to the proportion of the ingredients or the levels of fiber. It turns out that the fiber content of one such "feed" I used to buy on occasion ranges as high as 75 percent.

MILLING BY-PRODUCTS

There are all sorts of by-products from different milling processes. Some make excellent feeding supplements. Others aren't so hot. In general the word "feed" goes with lower protein by-products having fiber contents in the 20- to 30-percent range. The word "meal" goes with higher-protein, lower-fiber products. There are no government regulations that define these terms, though, so the safest bet is to buy feeds having stated values of protein and fiber.

If there is any question about the quality of a grain or mixture, whether it's been bought or raised on the farm, send a jam jar sample to the state or provincial agricultural testing laboratory for a complete analysis.

CHAPTER 10

Measuring Feed Ingredients

Most commercial milk farms run producing cows on what are called "full-feed" systems. If cows were gasoline motors we'd call it full throttle, and the analogy works. Commercial producing cows are fed all of the highest-value, most easily digestible food they can eat in relation to the milk and butterfat their genes will let them produce.

It's a tough grind and the toll is fierce. The average commercial dairy cow in North America only lasts about three-and-a-half years in production before being canned, either because she doesn't measure up to performance requirements or because some vital part blows up.

On the average, cows could produce more milk over their life-times if they were merely allowed more rest between lactations than present economics will afford. Present economics looks at that time and space taken up by one cow and says, "Gimme. Now!" It's going to take the farmer around 70 hours a year to milk a cow whether she gives 7,000 or 14,000 pounds of milk in return, and there it is. Fourteen thousand it's got to be, or out the door she has to go . . . and into a Family Cow barn at beef price.

The Cow's Needs

One principle that underlies all feeding methods, high-powered or low, that was hinted at in the last chapter is that a milking cow's food needs can be divided into four categories. These are her needs for growth (through her first lactation or until she has reached full size), for body maintenance, for milk and butterfat production, and for reproduction or gestation of a calf.

Each of the categories can be divided again into what is needed in the way of protein, energy, minerals and vitamins.

BODY MAINTENANCE

Body maintenance is the easiest and cheapest requirement to fill because the needs for proteins and minerals are lower than they are for the growth of bone and muscle or the production of milk. The needs for reproduction are barely noticeable in a cow until the last two or three months of gestation. Then they begin to intrude, but not enough to be of much concern to a family cow, especially if her milk production has dropped to around 20 pounds per day and she's headed for a lengthy dry period prior to calving.

It's a good thing that body maintenance is easy to fill because this takes priority over all other categories. A cow that is seriously underfed will stop growing and stop giving milk before she herself begins to waste away from lack of nutrition.

MILK AND BUTTERFAT

A "good" dairy cow puts foods that are above what she needs for maintenance into milk. A poor-quality dairy cow, like a cow from one of the beef breeds, is likely to put a greater share of those above-maintenance nutrients into excess body fat.

A number of specialized systems have been worked out for measuring and mixing the ingredients for total food rations for commercial cows. In most cases they aren't much use for the family cow because they depend on the farmer knowing fairly precisely what each cow needs from month to month, which takes knowing what each is giving in the way of milk and butterfat. I doubt that many family cow owners will bother with monthly lab tests. And even if they do there is the added fact that most commercial systems make allowances in their feeding recommendations for up to 5 percent of good nutrients that are wasted by cows on full-feeding programs—a fact of life, since cows' digestive systems work least efficiently at the upper limits of their production capabilities.

In spite of the drawbacks there are advantages in knowing some things about commercial systems and in using whatever ideas and symbols may help take the guesswork out of feeding one unpressured cow. People who want to go all the way with full-feeding should pick up a copy of *Feeds and Feeding, Abridged,* by Frank B. Morrison, which is listed under *Other Books and Places.*

For several years now, commercial feeding in Canada and the United States has been based on a system that measures a food's value on two scales. One is its *protein value,* and the other is its value in what is called *total digestible nutrients* (TDN).

Away from the nutrition laboratory, the proteins and related compounds can be thought of as the building materials for cow tissues, including milk, and TDN as a measure of a food's gross energy or fuel for building.

Total Digestible Nutrients (TDN)

TDN is a sum of the percentages of digestible carbohydrates, fats (times 2.25), and digestible proteins in a feed. The proteins are included in this gross energy measure because any excess in a food over what is needed for building materials can be broken down for carbohydrate-equivalents of energy.

It may be that within a few years not much will be heard of TDN. It is already being replaced in universities and many feed mills by renewed interest in an old system that measures a feed's net energy value.

The reason for the switch is that TDN can be off as a food value indicator because it doesn't take into account the time and energy spent in digestion. For instance, a high-fiber food costs more in the work of digestion—chewing, rumination and all its losses in heat, methane and carbon dioxide, and more—than does a food of equal nutrient value that's made of soft, easily dissolved parts.

Therms and Megacalories

Net energy values are hard to determine. It takes months of experimenting with cows in special chambers fitted with machines that measure many body processes. When they are all done, a food is given a

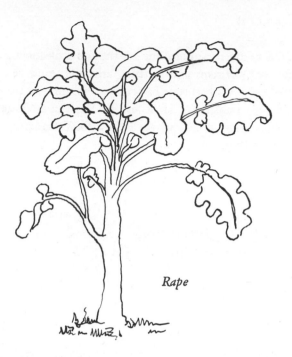

Rape

rating in million-calorie units called *Therms* (one calorie being the heat needed to raise the temperature of a gram of water one degree centigrade).

The rating also may be expressed in *megacalories* (Mcal), in which case they are talking about so many thousands of Calories. This Calorie with its capital "C" represents 1000 of the so-called "small" calories defined above.

Because working out net energy values is a long and expensive process, very few feeds have been given the full test. Usually the Therms said to be in a pound or 100 pounds of a food are *estimated net energy* (ENE) values.

After years of experimenting in laboratories and on farms across North America, tables have been drawn up listing the TDN and protein values of just about every type of food imaginable. The list for dry roughages alone in *Feeds and Feeding, Abridged* has nearly 300 entries between alfalfa and yucca. After each entry, values are also given for each food's percentage of dry matter, fiber and individual macronutrients. In many cases major minerals and vitamins A and D values are included.

Many years and experiments, too, have gone into figuring the daily requirements in protein, TDN, minerals and vitamins for the average cow at any age and any stage in milk or calf production. Several

121

feeding standards have been worked out on the basis of these experiments. In addition to Morrison's book there are tables and standards available through state and provincial agricultural services and through Canadian and U.S. National Research Councils. See *Other Books and Places*. NRC now has gone metric, which is fine for Canada, but a bit confusing for people in the United States.

Tables and Standards in This Book

Now there is still another feeding standard in the appendix of this book, along with a short list of feeds. Both of them are based on information from several sources that has been rounded and streamlined to provide guidelines only to feeding a family cow.

Understand that no one needs to use tables and feeding standards. For years past and years to come people have raised and will raise and keep cows perfectly well without a thought for "crude protein" or TDN. But there have been many times when I wished I had some rough framework to bounce around in—some general guide that would tell me approximately what the cow needed and how well I was filling that need.

The ideas behind and the methods for using the *Family Cow* tables and standards are exactly the same as for any commercial set. Only a measure of precision is gone because all values have been rounded off to the nearest tenth or whole number, and because I have only bothered with three weights of full-grown cows and have only provided figures for feeding for productions of 20, 30, and 40 pounds of milk daily.

I don't see that it matters. The precision that might lie in any other standard is lost or wasted anyway unless cows are being full-fed, have been accurately weighed, and are being regularly tested for milk and butterfat production.

In demonstrating how feeding standards and feed analysis tables are used—in this case the *Family Cow* models—I'll set up a simple winter or barn-season problem. From there it should be easy to expand their use to fit whatever peculiar situation comes up—or throw them out the window and read on to the next chapter.

FIGURING YOUR COW'S NEEDS

To use any standard, you want to know about how much the cow weighs and how much milk she is giving a day, as well as about what percent of butterfat is in the milk.

For this example I'm saying the cow weighs about 1,000 pounds, that she's giving 30 pounds a day, and that because she is a Jersey-Guernsey cross the butterfat is right around 5 percent.

The first thing to do is to use the *Family Cow* standard to add up this cow's total (maintenance plus production) requirements.

POUNDS

	Crude Protein	TDN	Ca	P
Maintenance	1.1	7.0	0.09	0.05
Production	2.0	11.0	0.12	0.09
Total	3.1	18.0	0.21	0.14

For the sake of this example I'm going to say that the cow is eating 20 pounds a day of a hay that most closely resembles what's described in the *Family Cow* feed analysis tables as "Timothy, good." Out of this hay she should be getting: 1.3 pounds of crude protein, 9.8 pounds of TDN, 0.07 pounds of calcium and 0.028 pounds of phosphorus. She's left with a deficit of 1.8 pounds of crude protein, 8.2 pounds of TDN, 0.14 pounds of calcium and 0.11 pounds of phosphorus.

To fill the deficit we could turn to a commercial formula feed or concentrate mixture. Good ones have TDN ratings of 70 to 75 percent. Since they are low in fiber (11 percent or less), their Therm ratings are right in line with TDN, being 70 to 75 in 100 pounds. Good concentrates also provide 0.6 to 1.0 percent of available calcium and 0.55 to 0.75 percent of usable phosphorus.

To figure how many pounds daily to feed of the concentrate mixture, divide the pounds of TDN required by the average TDN percentage for good feeds, say 73 percent (.73). In this case the answer is approximately 11 pounds.

Commercial concentrates may contain anywhere from 12 to 20 or more percent of crude protein. To find out which to feed, divide the

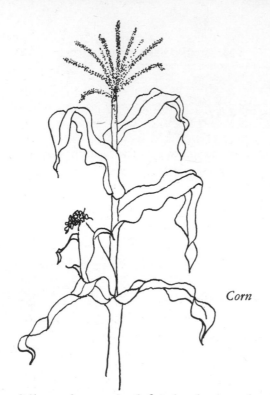

Corn

daily crude protein deficit by the pounds of concentrate that are to be fed. In this example a 16 percent CP concentrate will fill the bill.

If the concentrate mixture provides calcium and phosphorus in the upper range of what these feeds can provide, the cow's needs for these two minerals will be filled.

If the cow's ration left her shy in these minerals they could be fed as a supplement using one of the products suggested in the Appendix. Select a mineral supplement that supplies either calcium, phosphorus or both in a ration that is in line with the apparent need. In other words, if the ration leaves the cow shy twice as much calcium as phosphorus it wouldn't make sense to feed her a supplement like calcium phosphate, which provides the two minerals in very nearly equal amounts.

Pasture Values

The *Family Cow* feed analysis table offers one suggestion for the nutritive value of pastures: "Mixed grasses. 4 percent crude protein and 15 percent TDN." At first this seems a bit brief for a subject as important and full of variety as are the pasture forages. But this last is the point. Pastures vary greatly. Even one pasture can vary so greatly from one week to the next, as weather or growth processes change, that it is almost impossible to nail their values down with neat sets of numbers. For this reason all standards are least worthwhile for cows on pasture. You just can't tell what they're getting out there. This week the cow we were using for the feeding example might go on pasture and find young grasses so to her liking and needs that she'd eat lots and be able to produce those 30 pounds of milk without any concentrates whatever. Hooray! Cheap milk.

But don't let it fool you, because three or four days in a row of hot, dry weather could bring those lush pastures to bloom, seed and lignin. This means lowered nutrient value, lowered digestibility and lowered palatability. The only feed analysis table worth much at a time like this is in the milk pail. When commercial farmers put their cows out after a barn season they cut back gradually on hay and concentrates. They can keep on cutting back so long as milk production tells them that the pastures are taking up the slack. But let production level off (or worse—take a dip), then on they come again, building back with the concentrates and possibly other supplements as well, depending on individual circumstances.

Eventually comes a time in the northern temperate world when pastures, while perhaps still providing maintenance rations, can do no more. It becomes better all around to bring the cows in and feed them hay, other roughages or succulents having nutrient values that, with varying degrees of success, were locked in at the time of harvest.

CHAPTER 11

Grass/Hay and Concentrates—a Basic Diet

Here is a cookbook way to stuff a milk cow: no feed-value tables or feeding standards, no mention of TDN. But it's good; it works; and it offers a kind of step-wise timetable that's basic to all recipes for feeding cows.

Pasture Grasses

Let the cow have all the pasture she can graze, day and night and year 'round in warm climates, taking her off only if a freak storm hits or the ground becomes so wet in a rainy spell that her hooves are turning the grass sod to mud. Farther north, pasturing can begin each spring as soon as the new grass has pushed up four or five inches.

Most people let a cow run a couple of hours any fine day through the winter when it's convenient. There won't be much food value in it, but it is important that the cow get daily exercise, and sun if it is available.

If there is ever any question whether the cow is getting enough grass, put some hay in the pasture, either loose or (better) in some kind of manger that keeps the cow from spreading, trampling and wasting two-thirds of what she's given. Let a cow have some hay each morning the first few days she's on green pastures after the barn season. It will help her over the "Jersey squirts"—diarrhea that occurs partly because green grass is a wet and somewhat laxative food, and partly because the species and varieties of microorganisms that predominate in

126

a cow's rumen after months of barn feeding are not the types best suited to the digestion of lush pasture forages. There is likely to be a time lag of several days as the microbe population shifts to one better suited to the new foods.

Access to dry hay may prevent bloat, which sometimes happens when cows are in a pasture that is mostly lush, fast-growing legumes, especially alfalfa. Some people think the hay tickles and stimulates the rumination process and keeps cows belching so that they don't become bloated. There are other theories as well. Certainly a pound or two of dry hay never hurts a cow on green pastures.

Hay

When the weather gets nasty or for any reason it is certain that the pasture is no longer producing, give the cow all of the clean, un-musty and preferably leafy hay—as opposed to straw-like—she will eat. Fill up her rack or manger three or four times a day, the amount each time being judged by how well she cleaned up the last batch. Don't pile fresh hay on top of old. After a cow has drooled and fussed through a manger of hay for a few hours she has eaten what she wants, and the rest makes good bedding.

An old-time practice on farms where they were maintaining cows on hay was to feed exactly so many pounds a day to a cow. Around 25 pounds per cow per day is the figure I've heard most often. I don't know if these people weighed out the hay every day or if they just knew by the feel after dozens of years. Certainly if a 1000-pound cow could be coaxed into eating 25 pounds a day of even the lowest quality straw she would get along all right so far as her maintenance was concerned.

It may be interesting to know about how much hay the cow is eating a day, and here's a simple way to find out. Take a set of bathroom scales out to the barn and weigh yourself plus 30 pounds of hay. Feed through a day from that 30-pound pile. Then weigh yourself again with what is left in the pile plus all of the rejects cleaned out of the manger.

If it's found that the cow is eating less than two pounds of hay per day per hundredweight it's probably a poor-quality hay. Steps should be taken to increase the intake. Feed the hay more frequently or try making it more palatable with diluted molasses. If these steps don't help, it would probably pay to go out and buy a ton or two of better hay for supplemental feeding once a day.

Concentrates

Feed the cow five to ten pounds of a commercially prepared 16 percent CP concentrate each day, dividing the amount into two feedings, morning and night. The concentrate should go into a clean manger or be fed from a clean box or bucket that's kept for the purpose.

Normally this mixture of grains and protein supplements is fed throughout the year, except for two to four weeks beginning a day or two before the cow is "dried off" to rest up for her next calf. Start again with the concentrates about a week to ten days before calving. (See Chapter 16 for feeding heifers due to calve.) Concentrates should

A set of bathroom scales can be used to weigh hay to find out how much the family cow is eating in a day.

never be piled on suddenly. Begin with a pound the first day and gradually increase the level daily by half- or one-pound steps. The pre-calving
level should not go beyond 10 pounds daily.

FEEDS AT CALVING TIME

From a few hours before calving to a few after, the cow likely will
lose interest in concentrates. The first concentrate feeding after calving
may be half what the cow has been getting. Then over the next few
days she is gradually built back to her pre-calving level.

An older system for starting concentrates was to not begin until
after calving. Again, though, they started low and built by half- to one-
pound steps with each feeding. The entire "dry" period was also a fasting period for concentrates, with the exception perhaps of a half pound
or more of some laxative food like wheat bran just before the cow was
due to freshen.

The theory behind holding off on the concentrates until after
calving was that they caused the udder to swell. Now it is felt that
building a cow up to something like a ten-pounds-per-day level has no
effect on udder edema, and that in fact it is more reasonable to get the
cow started on concentrates before calving, since this is when her system is beginning to throw itself into milk production anyway.

MILK PRODUCTION VS. CONCENTRATE COST

Deciding how many pounds of concentrates to feed depends
mostly on how much milk is wanted, as measured against the cost of
the concentrates. (Tags on many commercial pre-mixed bags of concentrate suggest feeding a pound or so of the concentrate daily for every
three pounds of milk produced. This is a crude and possibly wasteful
rule of thumb, since it ignores so many variables.)

If a good dairy-type cow is on unlimited good pasture or good hay
her maintenance needs will be filled by these roughages and she will be
able to put all of her concentrates into milk. This is especially true during the first half to two-thirds of her lactation period. Depending on
the quality of the roughages, five pounds daily of concentrates might
sustain a yield of 20 to 30 pounds of milk daily from the small breeds

producing richer milk, and up to 40 pounds daily for the larger breeds producing the low-fat and non-fat-solids milk.

At 10 pounds daily, yields might be in the neighborhood of 35 pounds of milk for the smaller breeds and up to 50 with the larger animals.

Early in lactation a good milk cow probably would top those average levels of milk and fat production by drawing on her own reserves of fat and other body tissues in order to meet the demands. Even full-feeding can't always keep up with early-lactation demand in a cow genetically programmed to produce 50, 60 and more pounds of milk in a day. She meets the demand only by "milking off her back" until production drops back to what can be maintained by a daily intake of nutrients.

I heard of a man complaining about high-production cows, saying they were nothing but a fraud. "They give all the milk at once. After a couple months they drop right off to nothing." What may have happened was that the man came across some cows capable of 20 thousand pounds or more of milk in a year, but being used to smaller cows of lower potential he only fed them for seven or eight. As soon as the cows had burned up their reserves they could only crash.

HIGH-LEVEL CONCENTRATE FEEDING

There are times through the year when higher levels of concentrate feeding, approaching ten pounds daily, will be more helpful to milk production and the cow. These are (1) when the cow is on low-quality pastures or only fair-quality grass hays with no legumes; (2) during the first three months of lactation when milk production is at its highest; (3) when cows are giving very rich milk; (4) when cows are in their last two months of lactation—to help with developing the new calf; (5) when young cows are still growing; and (6) any time a cow is in a run-down condition.

In the feeding example in the last chapter it was figured that 11 pounds of a 16 percent CP concentrate were needed in addition to 20 pounds of hay to sustain 30 pounds daily of 5 percent milk. If that cow had only been getting 10 pounds of concentrates, her production might well have fallen three or four pounds a day. So it goes. This five to 10 pounds daily of concentrates isn't meant to produce record breakers. It's only meant to keep a family in milk and butter.

You certainly may feed more than 10 pounds of concentrates daily if you want more milk, if there is room in the cow for more dry matter, if she's eager to take more concentrates, and if she responds with enough more milk to make it all worth while.

RICHER MILK TAKES RICHER FEEDING
(Average amounts of dairy concentrates having a 75-percent TDN rating, needed for each pound of average milk produced by different breeds.)

Breed	Grain Mixture (Pound)	Digestible Nutrients (Pound)
Holstein-Friesian	0.41	0.307
Ayrshire, Brown Swiss	0.46	0.340
Guernsey	0.52	0.391
Jersey	0.56	0.419

(From USDA *Yearbook of Agriculture*, 1939.)

Just be sure when you get to feeding 15 or more pounds daily to watch carefully for signs of stress. Watch for indigestion and mastitis. The potential for these and other problems increases with higher levels of rich foods. It takes experience, at least, to pull it off.

It would be best to follow feed analysis tables and something like the *Family Cow* standard if you're going to start pushing the cow this way. The next best bet would be to follow the recommendations of a concentrate feeding schedule such as the one in the Appendix.

PROTEIN LEVELS

Another question of when a person might want to feed higher than 16 percent CP concentrates. It might be for a cow that was producing exceptionally rich milk, or when hay or pasture forages were of very low quality. If higher protein concentrates are not available it's usually easy and maybe cheaper to mix them at home by adding a high-protein supplement to the 16 percent CP feed. (See Chapter 12.)

Water

Make sure that the cow gets all the fresh water she wants. If she hasn't got a trough, stream or an automatic waterer, carry buckets of water to her three times a day. (You can get away with two waterings a day, but the more frequent access to fresh water the better.)

A cow may drink 10 to 15 gallons of water a day. The need for fresh water goes up with the heat of the day, with the nature of the feeds she is eating and with milk production.

Don't leave a cow alone with a bucket of water in her manger or within the circle of her tether. She is sure to knock it over as soon as your back is turned.

Minerals

The cow must have stock salt. If she has been without salt she may take too much to start, so begin a few days in advance by giving her a tablespoon of salt morning and night, on or mixed in her concentrates. This could be salt from your own table, although you will probably want to switch to a stock salt for the additional minerals that may be in it. After two or three days of measured salt, the cow may be given the salt free-choice, either granulated or in block form. Loose granular salt should be sheltered by a roof or box. Block salt doesn't have to be sheltered but it will last longer that way, especially in humid climates.

If there is any concern that the cow may not be getting an adequate supply of other minerals such as calcium and phosphorus—and it's more likely to be phosphorus that's deficient in the kind of high-roughage diet talked about here—she could be fed about a quarter to a half cup a day of steamed bone meal (which supplies both minerals) mixed into her concentrates. Or else a second salt supply could be set up, in this case a box with a mixture of two parts of bone meal to one part of granulated salt, by weight. There are about four cups of bone meal in a pound, and two to two-and-a-half cups of salt to the pound.

It could be a waste of money getting any deeper into minerals without first talking with a local veterinarian or extension agent.

If there is enough cobalt (for Vitamin B_{12}-synthesizing bacteria) in local soils or if it has been added to the stock salt, then the only vitamin deficiencies might be in A and D, and then only during the barn season since green feeds and sunshine provide quantities of both. A and D supplements are available at any feed store.

More About Hay

There are few places in Canada and the United States where it won't be necessary to put up at least a ton or two of hay each year to supplement pastures that are either frozen, too wet or too dried out to carry a cow. Even in the dry Southwest where pastures might provide a maintenance ration year-round, it still helps to have a ton or two of hay on hand to carry a cow through freak blizzards or other emergencies.

From midway United States north through agricultural Canada, three to three-and-a-half tons of good hay are suggested as necessary to

There is no tool like the scythe for harvesting hay or grains on rough, rocky land. It saves energy, too, but it's slow; and learning how to use a scythe properly takes weeks of practice. (Photo courtesy of Vermont Extension Service archives.)

keep a cow through the winter. Certainly the highest figure should be followed, plus. It never hurts to have extra hay. Stored in a dry barn, hay will keep well for more than a year. I've used loose hay that had been in a barn for five years. It was pliable, quite green and the cows loved it.

It is better to have a ton of hay to sell than one to buy in late spring when the market may be at its highest. It is also better, especially with lower-quality hay, to feed frequently and in plentiful quantities. That lets the cow pick out the best, much as she picks over a pasture.

RECOGNIZING HAY'S NUTRITIONAL QUALITY

Learning to estimate the nutritional quality of hay takes time and experience. Knowing the nutritional quality takes a laboratory analysis. Send a plastic bread bag sample of hay to a state or provincial agricultural department laboratory. Make the sample up from tufts taken at random from a mow or field, or out of several bales from a purchased ton.

Good hay smells fresh, not moldy. It has a green cast to it, and if you twist a handful it won't crumble and break apart like dried spaghetti. Moldy hay is dark and mushroomy smelling. Overmature or weathered hay is tan to brown. It may be full of seeds and will tend to crumble and be dusty. Good hay won't be full of coarse stems. Good legume hay is leafy. Red clover hay should have dried blossoms, not seed heads.

BUYING HAY LOCALLY

Buying baled hay can be tricky. The price of a ton may vary 20 to 30 dollars in the same valley at the same time of the year. The difference reflects quality and different people's ideas of how much a ton weighs. With string-tied bales weighing anywhere from 30 to 60 pounds each, and wired bales weighing maybe 100 or more, there can be no set numbers of bales to a ton. I made the mistake once of agreeing over the phone to buy a ton of hay from a man who claimed 45 of his bales made the measure. He delivered 45 loose bundles of junk, many of which weighed 30 pounds or less.

In buying from a local dealer it is often better to go by a price per bale with an understanding of how much the average bale is going to weigh.

One of the best and cheapest ways to buy baled hay is to pick it up locally out of a field. That way you have a better idea of the hay's quality, when it was made and whether or not it got rained on in the process. Hay that gets soaked after baling may look all right on the outside but be black in the middle.

Another good and cheap way to get hay is to trade your labor with a neighboring farmer in exchange for 200 to 300 bales.

CHAPTER 12

Shaving the Food Budget

Family cows don't necessarily bring in any dollars, and the fewer that have to go out in keeping them the better. From this angle there are a couple of possible drawbacks in the basic diet outlined in the last chapter.

This basic diet of roughages and concentrates relies heavily on commercial pre-mixed feeds. It also relies heavily on pastures and hay crops that can fail in an exceptionally wet or dry year, making it necessary to buy even more commercial concentrates or costly imported hay to fill in the gap.

There are three ways to shave the food budget: The first and simplest involves any trick or combination of tricks that will increase wet and dry roughage production or improve the quality of these roughages so that they become more nutritious and palatable. These are the cheapest foods. They are the easiest to produce, to harvest and to store on the farm. The goal should be to have them fill at least that part of the cow's ration that goes into body maintenance.

Once maintenance has been met, some high-carbohydrate, low-fiber foods—either home-grown or discarded by some nearby food-processing plant—may be fed in place of high-priced grains as part of the production side of the ration.

Finally, it may be possible to mix cheaper concentrates at home, using local grains or by mixing low-protein concentrates with purchased protein supplements to bring the whole up to the desired protein level.

Pasture and Field Management

In some southern states excellent year-round pastures, supplemented by good-quality dry roughages, have kept cows in high production without the addition of any concentrates.

The farther north or up mountains we go, the more difficult this becomes. But there always are ways to improve the roughage/succulent situation to get the most for what the climate will allow. These include improved pastures and pasture management, better hay crops and methods of harvesting, the planting of annual crops that are either grazed or cut green and carried to the cow, and the planting of annual roughages or succulents that can be harvested and stored for later feeding

FULL-SEASON PASTURAGE PLAN*

PASTURES	APR.	MAY	JUNE	JULY	AUG.	SEPT.	OCT.
Permanent							
Fertilized		▬	▬	▬		▬	▬
Unfertilized		▬	▬			▬	▬
Temporary							
Rye (Fall-seeded)	▬	▬	▬				▬
Oats			▬				
Supplemental							
Sweetclover (2nd year)		▬	▬	▬			
Sweetclover (1st year)						▬	▬
Kale or cabbages							
Sudan grass				▬	▬	▬	
Meadow (2nd crop)					▬		
Alfalfa (2nd crop)				▬	▬		

This scheme for combined use of permanent, temporary and supplementary pastures, to provide adequate pasturage throughout the season, originally was planned for Iowa. It could be used as a guide in other states, substituting temporary and supplemental pasture crops adapted to the locality.

(From USDA *Yearbook of Agriculture*, 1939.)

during the barn season as supplements to or replacements for part of
the hay.

ROTATE PASTURES

The simplest permanent pasture of a fenced acre or two of wild
grasses will give more value if it is split in half so that the cow can be
shifted back and forth at four- to six-week intervals. She tramples and
fouls less of the feed this way. Confined to a smaller area she grazes
more thoroughly, leaving fewer plants to grow over-mature and
unpalatable.

Rotating pastures lowers the pressure of overgrazing that tends
to stunt growth and that will eventually kill taller legumes. It also
makes it tough for those parasites that need, through larvae hatched
from eggs in the manure, to get back into the grazing cow. The longer
those larvae are forced to wait, the more will die.

RAKE AND MOW THE PASTURE

After a cow is taken off a pasture there will be a faster, more even
recovery if the area is raked over to break up and spread the cow flops.
Use a spring-toothed harrow, a section of chain-link wire fencing, a
dead tree or a string of old tires—whatever's handy.

When any pasture begins to look like a moth-eaten porcupine it
will help to clip it off eight or ten inches from the ground with a mowing machine. This keeps the tall, woody weeds from going to seed and
forces the grasses that were ready to go to seed to renew their growth
of more palatable and nutritious leaves.

INTRODUCE BETTER FORAGES

Domesticated varieties of pasture forages, that produce high-
quality feed over a longer period of time than native grasses, can be
introduced by plowing and reseeding or by a spring broadcasting of
seeds over land that was scarified with a disc or spike harrow. For

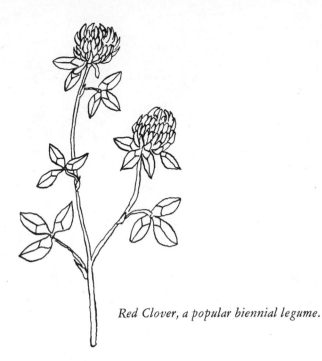

Red Clover, a popular biennial legume.

scarifying small pieces at a time, a rotary tiller will work where the grass is short and the sod is moist.

Untended hay fields revert to wild grasses and perennial weeds in a very few years. Tall wild grasses can make very good hay but they mature at different times and, over all, earlier than domestic grasses and legumes. Often they mature before the coming of the hot, dry weather needed for making hay.

Also the wild grasses may yield no more than a half a ton to the acre, whereas the domestic grasses and legumes on fertile, well-drained soils may give three or more tons of quality hay.

CUT HAY TWICE A YEAR

Cutting any hay field twice a year will increase the total annual yield and increase the nutrient value of the hay. The first cutting would likely be taken earlier to allow time for the growth of a second crop or rowen—roughly six to eight weeks. This earlier-cut hay would be less matured and weathered; lower in tough fibers and higher in protein value. Early cutting releases the slower-starting legumes that then have a chance to flourish before the next cropping.

139

PLANT GRASSES WITH LEGUMES

Although legumes have unequaled food value, they are usually seeded with grasses—for several reasons: Cows don't seem to like a strict diet of legumes—especially alfalfa—month in and month out. Legumes alone may create a lush, thick growth that is difficult to cure in humid regions or where people don't have equipment to crush or crimp hay as it's mowed, to speed the drying.

Grasses help protect the less hardy legumes that are slower to gain a foothold. They guarantee a better yield of something in case severe weather kills off a portion of the legumes. The grasses fill in a field as biennial or whatever shorter-term legumes die back. And the grasses are a hedge against the possibility of bloat that may occur with cattle pastured on lush, pure stands of actively growing legumes. For more notes on pasture and hay field improvement see Chapter 18.

Temporary Pastures

Large farms may have several acres of temporary pastures planted in annual forages that come into their own before or after the perennials that dominate the permanent pastures. The same could be done on a small scale with, say, the house garden being planted in fall to a winter rye that could be grazed off in the early spring. Or the new year's garden could go in a new spot while last year's was planted in an annual for late summer or fall feeding.

Many large farms maintain their cows throughout the growing season on crops that are chopped and fed green to the idlers waiting by the mangers. It's called "zero grazing," mentioned in Chapter 1, and something along the same line can be worked with the family cow. Years ago it was known as "soiling," and it was more a pasture supplement than a pasture replacement practice.

Soilage Feeds

Soiling crops or soilages were any crops, though usually annuals, that were cut green and hauled to the cows to take up the slack of fading pastures. There is no need to chop the feeds. They are chopped

NUTRIENTS IN YOUNG AND OLD PLANTS

The palatability of young plants almost invariably is superior to that of more mature plants. Timothy cut at the bloom stage is more palatable to dairy cows than hay cut at the seed stage. Young plants are tenderer; the proportion of leaves is higher; and the hay made from them is softer —all of which makes for greater palatability. The contents of protein and minerals are higher, and that of the less valuable crude fiber is lower in young plants than in older. These changes in the composition of alfalfa are shown in these tables.

THE EFFECT OF TIME OF CUTTING ON THE CHEMICAL COMPOSITION OF ALFALFA HAY. (Average for four years.)

Stage of Maturity	Ash (percent)	Protein (percent)	Fiber (percent)	Nitrogen-Free Extract (percent)	Fat (percent)
Prebloom	11.24	21.98	25.13	38.72	2.93
Initial bloom	10.52	20.03	25.75	40.67	3.03
One-tenth bloom	10.27	19.24	27.09	40.38	3.02
One-half bloom	10.69	18.84	28.12	39.45	2.90
Full bloom	9.36	18.13	30.82	38.70	2.99
Seed stage	7.33	14.06	36.61	39.61	2.39

COMPOSITION OF THE DRY SUBSTANCE OF TIMOTHY HAY HARVESTED AT DIFFERENT STAGES OF MATURITY

Stage of Maturity	Protein (percent)	Ether Extract (percent)	Crude Fiber (percent)	Ash (percent)	Nitrogen-Free Extract (percent)
1 foot high, no heads showing	10.18	4.61	26.31	8.41	50.49
Beginning to head	8.02	4.07	31.15	7.61	49.14
Full bloom	5.90	2.38	33.74	6.10	51.89
Out of bloom: seed found	5.27	3.13	31.95	5.54	54.12
Seed all in dough	5.06	2.87	30.21	5.38	56.48
Seed fully ripe	5.12	2.72	31.07	5.23	55.87

(From USDA *Yearbook of Agriculture*, 1939.)

commercially only because it's the quickest, "cheapest" way to handle these heavy feeds on a large scale.

Several family cow owners I have talked with plant an eighth- to a quarter-acre of land each spring to an annual that will be fed as a soilage in the late summer or fall. Last year a neighbor had a piece planted in oats and rape that he would scythe and feed to his cow, a wheelbarrow a day. He could have let the cow graze the oats and rape, but his way there was less tramplage.

Cabbages can be grazed or fed as a soilage crop.

CABBAGES, KALE

Cabbages are grown as a soiling crop in England and fed to milk cows at rates up to 60 pounds a day. Kale, too, is fed at this rate for fall and early winter, either being hauled to the cows or by letting them graze it off behind a movable electric fence. A tether could be used with the family cow. It is figured that a cow will graze about 60 pounds of kale in two hours. Rape, cabbages and kale are among crops that should be fed after milking, to avoid the possibility of their strong smells or flavors getting into the milk.

CEREAL GRAINS

Any cereal grain makes a fine soilage crop for mid- to late-summer feeding. Sweet corn stover (stalks with the ears removed) will be lower in value than any green grain cut earlier and carrying its seeds, but it will be welcomed by the cow and be of some help against increasingly low-grade pastures. Sudan grass is widely grown in corn country and southward as an annual hay crop or as a soilage. There is a danger of prussic acid poisoning with sudan if it's fed before the crop is about 20 inches high, and in the new growth just following cutting or freezing.

Leftovers

Many other foods are available at different times of the year; there are always leftovers from the garden. Any member of the cabbage family will be relished by the cow, a thing I discovered by dismal accident one afternoon when Gladys clomped once through the garden, methodically beheading 30 feet of broccoli. But think of the cow when you clean up old pea vines or bean stalks, over-mature beet tops, old lettuce or bruised pumpkins and squashes. Don't feed her onion tops or potato vines.

Most food processors leave edible and often valuable by-products behind. Some of these, like citrus and sugar beet pulps, now are most often marketed on a huge scale, and their prices approach those of grains. But if a person lives near a processing plant or cannery that hasn't established outlets for its leavings, there may be good cow foods available for the asking.

Depending on how wet they are, anywhere from three to seven pounds of these by-products would be roughly equal in energy (TDN) value to a pound of hay. Wet apple pomace from a cider mill or juice

Tip-toeing through the broccoli. Keep your garden fenced!

factory is a wonderfully high-energy, low-fiber food. Coming as it does just at the time of the year when pastures are pooped, makes me wish I lived next to a mill.

Silages

Commercial milk farms outside of ideal alfalfa-growing and curing country have been going heavily into corn, sorghum, and grass silages in recent years. They make wonderful feeds that can be put up in wet years that would drive haymakers crazy, but I don't think they work for a farm with less than half a dozen cows.

Silages are made by packing green-chopped or long-cut forages into upright silos or trenches covered with plastic (usually), where they are allowed to ferment in their own juices. Tight packing and covering are important, because if warm air gets at the silage the fermenting bacteria give way to putrifying organisms and the feed rots. Often when a standard silo is first opened to begin feeding, several feet of spoiled silage has to be thrown away—representing how far air managed to percolate into the mass.

In drier climates, whole corn for winter feeding may be left standing in shocks in the field.

From then on two to four inches of silage has to be fed off the top every day to keep ahead of spoilage—the warmer the weather the more has to be used. It doesn't seem to me that this could be handily done with only one or two animals. Another problem can be with silage freezing in very cold weather faster than it can be fed. Picking through frozen silage is an awful chore.

But there are alternatives to silage for winter feeding. Although lacking the juices and palatability of silages, there would be tremendous hay-equivalent value at least in field corn or one of the grain sorghums that could be grown and stored in the barn or, in drier areas, be left standing in shocks in the field for feeding through the winter. They could be fed whole or they could be run through a composter-chopper with large screen holes as you went along. Cows are likely to clean up chopped fodder more thoroughly.

Root Crops

After 30 years of being ridiculed, root crops are beginning to regain popularity in commercial milk farming, probably because of the high cost of grains. The growing of roots "certainly merits reconsideration on all livestock farms," says a 1974 agricultural bulletin from England. More recently the New Brunswick Department of Agriculture published recommendations for feeding potatoes to milk cows.

They made it sound like a new discovery, which must have had some of the old-timers scratching their heads, because before the advent of relatively inexpensive grains, silos and silage-making equipment, there was just about no other northern way to make winter milk other than with plenty of roots.

But with silage came new thinking, and in a lot of cases a determined effort to wipe out the old. An Ontario Ministry of Agriculture person asked about the possibility of feeding roots in the northern, above-corn reaches of that province replied, "too much labour to grow a crop that is over 90 percent water—we can pump water from a well!"

Well, wet they are. But mangels, the wettest of them all, have a lot more going for them. They yield up to 40 tons to the acre, and in that yield there can be as much or more dry matter than in an acre of corn made into silage. There will be more TDN value in the corn, but more fiber too, so more energy will be expended by the cow digesting the silage.

Cows love mangels and many other roots, and they do well on them. On a small scale, without mechanization, they are as easily grown and harvested as corn. They store easily and well in cellars or in shallow, covered pits called "clamps." (See Chapter 19.) In milder climates turnips, carrots, parsnips and other roots may be left in the ground until they're wanted in the feed trough. In the United States South, sweet potatoes and other roots can be sliced, sun-dried and fed—on an energy basis—pound-for-pound in place of grains.

Roots and other succulents are usually fed at rates of 30 to 60 pounds a day. Since almost all of them are low in protein they are used most simply to replace part of the hay or other dry roughages filling the cow's lower-protein maintenance needs. When they are used to fill production needs they have to be bolstered with high-protein supplements.

The accompanying table shows the average number of pounds of various succulents needed to replace seven pounds of good hay. Don't replace the first 10 pounds of long-fibered hay the cow needs for dry bulk and fatty acids. (See Chapter 8 on chemistry and nutritional processes.) The table should only be used as an approximate guide to equivalent values based on each feed's average percent of dry matter, TDN value and behavior in feeding experiments.

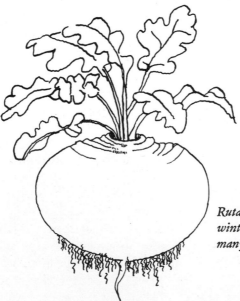

Rutabagas were once a popular winter feed for cows, as were many other roots.

HAY CAN BE REPLACED

A COMPARISON OF VARIOUS BULKY FEEDS
WITH AVERAGE HAY ON AN ENERGY BASIS

7 lbs. average hay = 20 lbs. corn (or sorghum) silage = 25 lbs. kale = 34 lbs. cabbages = 15 lbs. sugar beets = 26 lbs. sugar beet tops = 30 lbs. rutabagas = 50 lbs. turnips = 36 lbs. mangels = 25 lbs. fodder beet = 16 lbs. pressed beet pulp = 14 lbs. potatoes = 25 lbs. carrots = 18 lbs. brewers' wet grains = 3 lbs. barley.

(From "Advisory Leaflet 524," Ministry of Agriculture, Fisheries and Food. Pinner, Middlesex, England, 1974.)

Protein Supplements

Supplementing succulents with protein shouldn't be a problem, because there are several high-protein milling by-products on the market. Many combinations of succulents and protein supplements will be possible.

Old-time agricultural books often mention feeding 30 or 40 pounds of roots daily supplemented with a pound or two of some high-protein oil meal cake. This would be in addition to 20 or more pounds of hay or hay and straw. Sometimes they'd chop up the straw and roots and mix them together in a pile that would be left to ferment for several hours before feeding. Other farmers might cook or steam the mixture. In some cases these efforts might have paid off by increasing the palatability of the roughages. It's not likely that digestibility was improved.

Once maintenance requirements are filled with hay or hay and another roughage or succulent, 30 pounds of kale, cabbages, roots, fruits or fruit pomaces, beet tops, pumpkins, squashes or wet vegetable canning wastes, plus two pounds of one of the high-protein oil meals having 35 to 40 percent of crude protein, would be worth somewhere between five and ten pounds of a 16 percent CP concentrate.

Any more accurate figuring would take using the *Family Cow* feed value tables in the appendix. For example:

	CP lbs.	TDN lbs.
30 lbs. potatoes	0.6	5.25
2 lbs. linseed oil meal	0.7	1.46
Total	1.3	6.71
8 lbs. of 16 percent CP concentrate having 73 percent of TDN	1.28	5.8

Limiting Succulents

If a cow is already being fed around 30 pounds of a succulent toward filling her maintenance requirements, I would be more cautious about feeding more succulents for milk production. For one thing most roots are low in all minerals. Cabbages are too. Kale is low in phosphorus. So are the root tops. Fruits and their pomaces are low in calcium and phosphorus.

Some of the succulents may cause diarrhea if they are fed to the cow too fast or in larger quantities than she can handle. A cow may not like white potatoes to begin with and it may be necessary to mix them up with some sweetened concentrates.

The smell or flavor of some of the succulents may taint the milk, and so it's a good idea to get into the habit of feeding them just *after* milking. It would also be best to divide 30 or more pounds of succulents into two feedings, morning and night.

Roots and fruits sometimes can get stuck in a cow's throat, causing her to choke and maybe to become bloated because she can no longer belch. Some say there's no problem if the roots are large enough that the cow is forced to chew them up a bit. Others say there's no problem if the roots are fed from ground level. But, to be sure, all roots and fruits should be sliced, chopped or smashed before feeding.

Old-fashioned hand-cranked slicers and choppers sometimes can be found (shake out the dirt and geraniums). Or maybe one of those composters would do if it were fitted with a screen with inch-diameter holes or larger.

Green potatoes and potato sprouts are poisonous. You can rub the sprouts off and turn green potatoes white again by storing them in total darkness. Turnips store least well of all roots. Mangels should be dug and stored six weeks or more before being fed to prevent indigestion.

Most roots go about 60 pounds to the bushel. Figuring a maximum rate of feeding for 200 days, it would take about 250 cubic feet of storage space to put them away.

Making Your Own Concentrates

It's not easy to mix grain-based concentrates for less than they'll cost at the store. By watching the market, buying in bulk, and using chemical nitrogen in place of proteins, the mills are able to sell total concentrates complete with extra vitamins and minerals for less than they'll sell an equal weight of a single grain.

If just one good grain like corn, oats, barley, wheat or rye can be grown or bought directly and inexpensively from a producer, then it may be possible to beat the system by mixing the one with other store-bought grains and a protein fortifier.

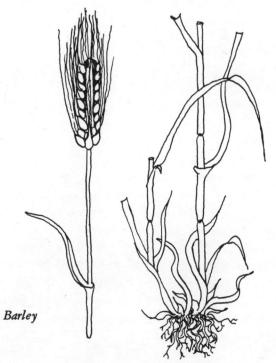

Barley

Concentrates could be made with one grain and a high-protein supplement, but usually at least two grains are included in the mix. This is to make sure that there aren't any problems with palatability or with possible bad effects to health or to milk and butter quality that might come from feeding too much of a single grain.

SUGGESTIONS FOR GRAIN MIXTURES

GRAIN MIXTURES HAVING DIFFERENT PROTEIN CONTENTS, TO BE FED WITH DIFFERENT ROUGHAGES

Roughage	Approximate total protein content desired in grain mixture (percent)	Ground corn (pounds)	Ground oats (pounds)	Wheat bran (pounds)	Cotton-seed meal (pounds)
Legume hay alone. (If clover, add 100 lb. cottonseed meal to grain mixture.)	12	400	200	200	—
Legume hay & silage or mixed hay (half grass, half legume)	16	300	200	200	100
Mixed hay (half grass, half legume) and silage	20	200	200	200	200
Grass hay & silage, or either alone	24	100	200	200	300

(From USDA *Yearbook of Agriculture,* 1939.)

For instance, 100 pounds of a 16 percent CP concentrate could be made out of a mixture of roughly 70 pounds of ground corn and 30 of ground soybeans. But a concentrate that's more than 25 percent soybeans may cause a cow to begin producing butterfat that's low in vitamin A value and that makes a soft butter.

Harvesting grain the old way. Before the invention of combines, grain harvesting began with cutting and tying the stalks into bundles called sheaves. *Sheaves were gathered into self-supporting* shocks, shooks *or* stooks *like the one pictured below. Later, when the grains were dry enough for safe storage the shocks would be gathered for threshing with flails or mechanical threshers. (Old photo courtesy of Vermont Extension Service archives.)*

USING THE PEARSON SQUARE

To work out what proportions of different grains and high-protein supplements to use in a concentrate mixture, most small operators use what is called the "Pearson Square" method. As an example of how this works, say a person had some barley and some cottonseed oil meal and wanted to make up 100 pounds of a 16 percent CP feed. The wanted protein level goes in the middle of a square, and the CP levels of the two ingredients go on the left-hand corners.

Subtractions are made across the diagonals of the square, with the remainders being placed on the right-hand corners.

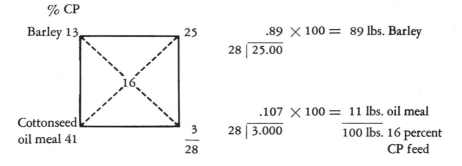

% CP

$$.89 \times 100 = 89 \text{ lbs. Barley}$$
$$28\overline{)25.00}$$

$$.107 \times 100 = 11 \text{ lbs. oil meal}$$
$$28\overline{)3.000}$$

100 lbs. 16 percent CP feed

The sum of the right-hand figures is the total parts in the feed. The pounds of each ingredient that would go into 100 pounds of the feed is figured by dividing this total into each remainder and multiplying by 100.

Two or three grains could be combined in whatever proportions were possible or desired. Then, using the Square method, the mixture could be brought up to a higher CP level by combining it with a high-protein supplement.

To find the CP percent of a random mixture of grains, multiply the quantity of each ingredient by its own CP percent. Add up the total pounds of crude protein and divide this by the total weight of the mixture.

A pound-and-a-half of bone meal and two pounds of a vitamin supplement that included A, D and possibly E could be added to each 100 pounds of a home-mixed concentrate, especially for winter feeding.

Sometimes local feed stores are out of—or never carry—higher than 16 or 18 percent CP pre-mixed concentrates. Using the Square method these lower CP feeds can be mixed with a high-protein supplement to build whatever percentage of crude protein concentrate is needed.

Even if the higher CP pre-mixed concentrates *are* available at the store, there may be a cash saving in mixing your own from one of the lower percentage feeds.

If the price were right, fish meals or meat tankage could be used as the protein sources for raising the CP level of a concentrate three or four percentage points. These products from rendering plants are in the 50 to 60 percent CP range. Fish meals and meat scraps that contain bone also will be rich in calcium and phosphorus. Although some feeders are experimenting with whole fish as protein supplements for beef cattle, a general problem with these supplements of animal origin can be a lack of palatability. Try a little before buying a lot.

A clean, crack-free barn floor is a good place for mixing small quantities of feeds made from several ingredients. Spread each ingredient separately over an entire area of floor, one on top of the other, and with the smaller portions going on last. Then use a flat, snow-type shovel to pick up the mix carefully section-by-section, transferring it to a hopper or a metal or plastic garbage can.

Always store grains covered to keep out rats and mice, and preferably keep the concentrates in a room separate from the livestock in case somebody gets loose in the night.

CHAPTER 13

Milk and Milking

The idea of hand-milking can be a great fear for people thinking about buying a cow. I know it was for me. I remember working on farms that were going crazy with mastitis, and hired men being bawled out for leaving milking machines on too long or for not taking the time to "strip" the cows properly. All the while I'd be running around forking silage in one end of the cows and manure out the other and finding it easy to think that the best thing that could happen would be to never be asked to milk them.

Certainly there was no hand-milking being done on any of the farms where I worked. And so in this respect, as in a lot of others, working on a commercial milk farm is not the best place to pick up the skills that go into keeping one or two cows.

With reluctance I had to tell the man who sold me my first cow that I did not know how to milk. I expected he would be somewhat amazed, maybe even a little disgusted. But he only laughed with good humor. "Go on," he said, "take her home. You'll learn fast enough."

Hand-Milking

And that is all there was to it. No, I didn't get all the milk that first night—or the next morning either. But I remember three or four days later running to the house with a bucket of milk that actually had a crest of foam to tell what a mighty milker I'd become.

When a person first takes a crack at milking it can be awfully

Hand milking is one of those agricultural skills that's hard to pick up on a modern automated farm.

frustrating. A squeak, a squirt, then nothing. Then a good squirt but it's down your arm. Your arms ache, the cow sighs and shifts to her other hip and—Oh Lord, but it's getting late. There's the eight o'clock news!

This is one of the reasons why I think it is better to buy a cow on the last leg of her lactation rather than one that's early in lactation and swelling with gallons of milk. It won't matter if the bag is left holding some milk.

With a little effort and patience almost every cow learns to stand quietly while she is being milked. Some will stand without being tied or restrained in any way, which is ideal for those who prefer to milk their cows in the pasture during fine weather.

For convenience and also to encourage cows to come at milking time, this is a good time to give cows whatever concentrates are being thrown into their daily ration. Some people dish out the concentrates before milking. Others wait until after milking, feeling that this way the cow stands more quietly because she isn't chasing grains around her manger. To the cow it only matters that the same sequence happens at each milking.

A milking stool can be anything that is easy to handle, comfortable to sit on, and low enough to get the milker's arms comfortably under the cow. Actually I gave up on stools after the last box I was using got laced with manure as the cow was leaving the barn. With only one cow I found it easy enough to squat beside her through the milking. Much more important, especially in winter milking when the barned-up cow may have a soggy tail, is a string to tie that tail out of the way.

One old-style milking stool that pro-vides a step for the milk pail.

On rare occasions it may be necessary to begin milking a day or two before the cow calves. This is only when the milk is fairly squirting out of its own accord, and only a little should be taken to relieve the strain.

Usually milking begins a morning or evening after the calf is born. The udder will be tight and swollen and yet in that first milking the yield will be low. With milking and massaging, the udder will become more pliable, tissue swelling will give way to milk production, and the same-sized udder will begin to yield a lot more milk. From then on the cow is typically milked twice a day (at regular ten- to twelve-hour intervals) for up to about 300 days, maybe more. It depends on how long the yield supports the effort and extra feed.

When to Milk?

When the cow is being milked twice a day, the times could be, say, 6 A.M. and 5 P.M., or maybe 10 A.M. and 11 P.M. But *not* 6 and 5 one day and 10 A.M. the next. Choose your own most convenient schedule and try to stick to it as closely as possible, realizing that if you goof now and then and miss the right time by an hour or two you won't be killing the cow, but neither will you be helping her or her milk production.

This 300-day, twice-a-day business is not an unalterable sentence handed down by some high court. The important thing for the cow's health and well-being is that she be kept on a regular schedule.

One year, after a couple of hundred days of milking, I switched in the course of three days to only one milking a day. Horrors! People said she'd dry up. Not a bit. She gave less milk, but the daily yield was only down about a third, and she kept on producing through 300 days. This isn't a recommendation. Another cow might dry up. But I have talked with other people who had my same luck.

If, at some point through the lactation, something comes up that interferes with milking, cut back or off on the concentrates and let the calf run with the cow. If the calf scours, take her off the cow for a day and then let them run together by increasing numbers of hours a day until they have adjusted to each other. If the cow seems to be swelling, milk her out a bit for a couple of days.

If the cow is producing lots of milk and there are other calves around, see if she'll accept the role of a nurse cow. Heavy producers can take on up to four calves, but I would leave it at two for the small family cow on limited concentrates.

The Udder

A cow's four teats are outlets for four separate "quarters" of the udder. The two rear quarters are heavier producers—usually about 60 percent of the total yield. The milk is produced in minute pockets called *alveoli* that drain into cisterns above each teat canal. Almost all of the milk a cow is going to give at a milking is in the alveoli and ducts leading to the cisterns when milking begins. The alveoli have been making milk over the past hours from ingredients taken from the cow's bloodstream. At a certain point the pressure of the milk within each little manufacturing cell becomes so great milk production ceases. For this reason it's possible to get more out of a cow through three milkings daily. This is done on some commercial farms, though twice-daily milking is by far the more common practice.

The alveoli actually push milk out when milking begins. It's an involuntary response triggered by a hormone from the pituitary gland that in turn gets its signal to flow from things the cow sees, hears or feels—maybe all three at once. The hormone continues to flow and the alveoli to contract for up to 10 minutes—it's variable with different cows—and then it's done. You can milk till your arm drops off and nothing much will happen.

THE "LET-DOWN" REACTION

People were aware of some "let-down" process long before hormones were known. All kinds of gimmicks and props and rituals have been used to bring it on strong. For thousands of years it was suggested the calf had to be present at milking time. If you didn't have a live calf a stuffed one might do. Or the milker would don a calf skin.

Then there was (and still is in some parts of the world) a practice of *insuflation* to encourage let-down. While one person milked another blew air or water into the cow's rectum. This was practiced in parts of Europe well into this century. Other cultures felt that mas-

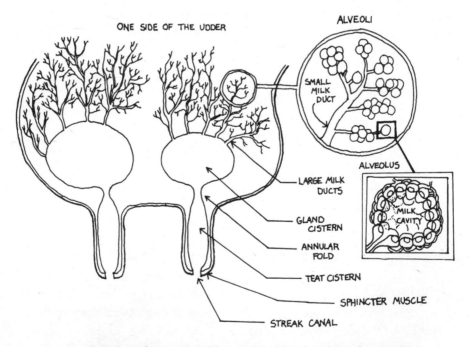

Interior of a cow's udder. Milk is produced in the alveoli *which, when stimulated by pituitary hormones released into the bloodstream at the onset of milking, contract, forcing the milk into the ducts leading to the gland and teat cisterns. (Adapted from California 4H bulletin.)*

saging the cow's vulva and anus was the best way to encourage the milk's flow.

None of this is as far-fetched as it might seem. Hormones closely related to the oxytocin that triggers milk let-down are important for other involuntary reactions at the back end of a cow. And it may well be that the wilder, less genetically refined cows of other cultures and ages do or did need this kind of strong stimulation.

The cows we are dealing with usually are so ready to be milked by almost any means that some will begin leaking milk as soon as they're called to the barn or hear the clatter of a pail around the usual milking time. Our problem is more to move right along when milking begins, to get the most before the let-down reaction has come and gone.

How to Milk Your Cow

First, wash the udder with warm water that's had a bit of disinfectant added. A small shot of household chlorine bleach is fine. This gets the let-down reaction going if it hasn't already started. It cleans away crud that otherwise would fall in the milk, and helps soften possibly bruised or scratched teats. (In this same vein don't milk with long fingernails.)

Then grab a stool or hunker down close into the cow's right flank. (It could be left or even from the rear, but right flank is what the cow is likely used to and what she'll probably get on the next farm.)

Cows can be milked one teat at a time, but two at a time is better for speed and for taking advantage of let-down. Any two teats can be milked at a time. Front two, back two, far side—whatever is least awkward.

Ease into milking. The cow may have a sore teat, and a sudden latching on may result in a pail full of dirty hoof. Hold the teat firmly and close over the top of the canal by pressing the teat between thumb and index finger. Then close the rest of your fingers while pulling gently down. *Do not crush!* Imagine that you are holding a tube of toothpaste that's sprung a leak in its bottom, and that you'd like to try squeezing it in one hand in a way that makes the paste go through the nozzle. If your thumb and finger aren't firmly together the toothpaste will backfire. So it goes in milking, often backing up into the cistern unless the top of the canal is closed.

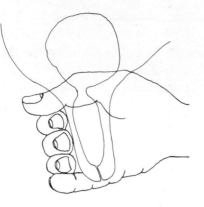

A

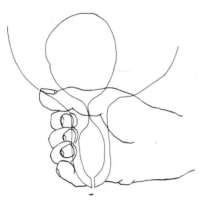
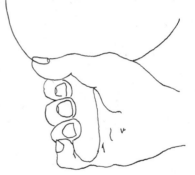

B

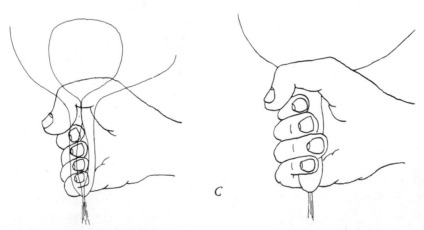

C

*In hand milking, (A) Grasp the teat high so that the thumb
and index finger are able to circle the teat where it meets
the body of the udder. (B) Close off the top of the teat
cistern by squeezing thumb and index finger together over
the annular fold. (C) Close rest of fist so that milk trapped
within the teat cistern is forced through the streak canal.*

If the milk flows or not, relax and repeat the process so that the canal can refill with milk. I've seen anxious new milkers haul down on the teat for dear life as though somehow a tighter, harder yank would eke out what certainly is no longer there. When two teats are being milked simultaneously, alternate between the two in this closing and gentle downward pull.

At the end of milking the quarters are "stripped" by massaging each in turn with one hand while milking out this last and richest milk with the other.

For the most part the potential amount and kind of milk a cow can give is in her genes. But there will be variations through a single milking, through a lactation period and through her lifetime.

Colostrum

The most radical change in quality is at the start of each lactation, as the colostrum becomes normal milk in a matter of three or four days. Colostrum is yellow and has the consistency of light cream. There may be a little blood in the colostrum the first milking or two. This is nothing to worry about. It's only because the tissues of the udder were under pretty ferocious strain up to calving and a few tiny blood vessels ruptured.

Colostrum is also called "first milk" or "calf's milk." It has nearly five times the protein of regular milk and is rich in vitamins and those antibodies the calf needs to resist colds and infections.

Most people turn up their noses at the idea of using colostrum in any way for human food. Maybe this is because when it is raw it can have a bitter taste. But cooking takes the bitterness away. Cooking also causes the colostrum to thicken, making it an excellent food for no-egg custards. (See recipe for Calf's Milk Custard, Chapter 14.)

The usual practice with colostrum is to give some to the calf and the rest to chickens and pigs. It can also be frozen and fed to the calf later on, or it can go on the garden or compost heap.

The colostrum will have given way to regular milk by the sixth milking or so. You can tell by the look. Most people wait until the eighth or tenth milking before taking a pitcher to their own table. Through the rest of a lactation period the most noticeable differences will be in overall volume of milk and in the butterfat level.

THE TRANSITION FROM COLOSTRUM TO NORMAL MILK

CHANGES IN COMPOSITION

Time After Calving	Total Solids (%)	Ash (%)	Protein (%)	Fat (%)	Lactose (%)
At parturition	27.42	1.37	13.97	8.45	3.63
6 hours	27.47	1.07	9.34	13.02	4.04
12 hours	15.63	0.89	4.77	5.68	4.29
18 hours	14.56	0.87	4.25	5.26	4.18
24 hours	13.98	0.87	3.99	4.88	4.24
30 hours	13.41	0.87	4.09	3.88	4.57
36 hours	13.54	0.86	3.85	4.08	4.75
44 hours	13.52	0.85	3.57	4.25	4.85
52 hours	13.35	0.86	3.66	4.14	4.69
60 hours	14.22	0.84	3.70	5.02	4.66
68 hours	14.17	0.84	3.79	5.19	4.35
76 hours	13.82	0.85	3.86	4.68	4.43
84 hours	14.70	0.81	3.58	6.79	3.52
11 days	12.78	0.75	2.92	4.33	4.78

Through one milking the butterfat level rises steadily. A cow giving 5 percent fat milk may give milk that is only a percent or two in fat with the first squirts. The last, the "strippings," will be up around 10 percent or higher. (Squirt directly into one cup of hot coffee for outstanding café au lait.)

VARIATIONS IN FAT CONTENT OF MILK DURING MILKING

Portion of Milking Period	Cow No. 1 (percent)	Cow No. 2 (percent)	Cow No. 3 (percent)
First	0.90	1.60	1.60
Second	2.60	3.20	3.25
Third	5.35	4.10	5.00
Fourth (strippings)	9.80	8.10	8.30

(From USDA *Yearbook of Agriculture*, 1939.)

Milk Production

Milk production climbs fast the first three weeks after calving. Then it begins to level toward a peak that will come somewhere between the fifth and eighth weeks. Full-fed cows often will maintain near-peak production for another five weeks and then begin dropping off by about 10 percent a month. Cows that aren't pushed usually will peak a little earlier than full-fed cows—lower, of course—and will slide off more rapidly. A full-fed cow's daily yield at her peak of production, multiplied by 200, gives a rough approximation of what she can produce over an entire 305-day lactation period.

A cow's annual production may climb 20 percent through her first two lactations. It will continue to rise more slowly through her sixth or

LACTATION PERIOD*

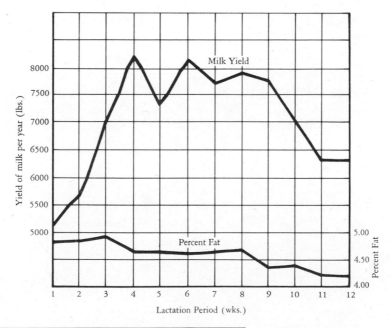

* Influence of age on the yield of milk and percentage of fat. Averages for six Jersey cows for 12 years.

(From *Dairy Farming* by Warren Eckles, Macmillan, 1916.)

seventh lactation, meaning she will be about eight or nine years old. From then on the yield begins to slacken off, but not drastically.

Anything that upsets the cow may lower milk production. She likes a routine and will be upset if it is radically changed—by milking several hours off the regular time, for example. It can be done—in fact a milking could be dropped altogether with a cow beyond peak production. But people who can't manage to give a cow a regular schedule will get lower production and run a greater risk of health problems.

Noise or confusion in a barn that is usually quiet may cause a cow to hold up her milk. If someone forgot to water the cow she will be down in milk. If she has a bellyache or worse she may be down. And if she is in heat she may all but cancel a milking.

If the cow suddenly drops off in production by obvious quantities and doesn't bounce back by the next milking or two, maybe even giving more to make up for the loss, then there could be something seriously wrong.

In most cases a poor diet, whether it is low in nutrients overall or only badly out of balance, will tend to lower the total output of milk before it causes any change in the milk's quality. There are exceptions, like the low butterfat in the milk of cows on low-fiber diets. Also a cow may continue to put out her normal yield of milk even when her diet— and so her milk—are so low in vitamin D, A or A's precursor, carotene, that a calf being raised on the milk becomes sick. But this wouldn't likely happen until a cow had been locked in a barn and fed poor grass hay for several months.

BUTTERFAT CONTENT

Generally, the more milk a cow produces, the lower the percentage of butterfat. Highest-yielding Holsteins give the lowest percentages of fat. It's said some smart Jersey farmers keep a Holstein around to milk last. That way they don't have to wash out the milking system! And then there are, for sure, many Holstein farmers who keep a low-yielding, high-fat Jersey on the place for their own fresh milk and cream.

The percentage of butterfat drops at first, then tends to climb a bit through a lactation as the cow's overall production drops off. The percentage of fat in winter may be higher than in summer. Some say it's

DAYS IN MILK
LACTATION PERIOD*

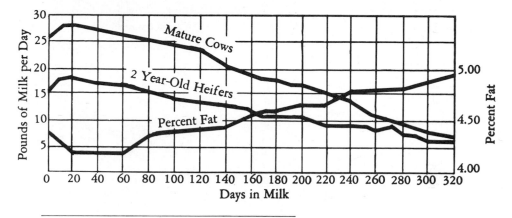

* Influence of the advance in lactation period on the milk yield and fat content. Averages for ten mature cows and ten two-year-old heifers.

(From *Dairy Farming* by Warren Eckles, Macmillan, 1916.)

due to the weather. With the family cow it might be higher when milk yield is down because the cow is on less than ideal feeds. (See Chapter 4.)

Butterfat is in droplets in the milk. The larger the droplets the more quickly they combine in churning to make butter. The cream of Jersey and Guernsey and probably Canadian cows has a higher percentage of larger fat droplets than is found in the cream from the other breeds. In all cows the percentage of larger droplets falls off as a lactation period goes along and as the cow ages.

Most of the color of milk and cream comes from carotenes the cow took from her food. One reason the milk and cream of Holsteins, Ayrshires, Shorthorns and Brown Swiss cows lacks color is because they convert more of their carotenes into colorless vitamin A than do the Jerseys or Guernseys.

LACTATION PERIOD ENDS

Within reason and with persistent milkers, it can safely be said that the lactation period ends when you decide. It comes when the cow is no longer giving enough to make milking worthwhile or when, as sometimes happens with cows giving 20 pounds or less a day late in lactation, the milk becomes bitter or salty. (See accompanying table, Off-Flavors in Milk.) It could come when some other commitment gets in the way and there isn't a calf around to put on the cow. Or it could occur when the cow is due to freshen in another eight weeks or so, in which case she must have a rest.

Drying Off

Three of the things that cause a cow to stop giving milk are used to dry her off: Lack of food, lack of demand for her milk and increased udder pressure. There are different theories about how best to spring the combination. Some say fast, others slow. In part it depends on how much milk the cow is giving at the time.

Slow or fast, the first step is always to stop feeding concentrates to the cow a day or two before drying her off.

OFF-FLAVORS IN MILK
(and some ways to avoid or correct for them)

Barny flavor. Time to clean out the barn and improve the ventilation.

Bitter milk. May be caused by feeds or weeds (see below), or be due to a high content of lipase which sometimes occurs late in lactation.

Salty milk. May be sign of mastitis or it may be from a high content of chlorine, also something that can happen late in lactation.

Miscellaneous feed or weed flavors. The list of culprits includes silages, cabbage, turnips, garlic, ragweed, wild onion and leek. Also quantities of fresh alfalfa clovers or grasses may impart a flavor if they are fed just before milking. These weed and feed flavors may get into the milk through what a cow eats or even what she smells if the odors are heavy in the air. If weeds are causing your problems, try to get rid of them or bring the cow in off the pasture where they are thriving at least four hours before milking. Apparently no off flavor will persist within the cow or her milk beyond four hours. Silage, cabbages, turnips and other known offenders are routinely fed just after a milking. That way there is no chance that their flavors or smells will taint the next milking.

Certainly if the cow is giving 10 pounds or less of milk a day the milking can be stopped abruptly. She will continue to produce milk for a day or so, then quit. The milk that is in her udder will be reabsorbed.

If the cow is giving 10 to 20 or more pounds a day, my own feeling is that it is better to dry the cow off over two or three days, especially if she has ever had infections in her udder. The first day after the concentrates are stopped don't milk her out completely. The next day take some milk in the morning and skip the evening milking. The third day sleep in. That evening a little milk could be taken if the cow's udder seems tight and swollen.

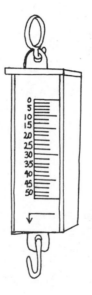

Inexpensive spring scale for weighing milk in the barn.

Keeping Records

It often is helpful to have some kind of running record of milk production, if for nothing more than to look back on in following years to see if there have been changes. Who knows what comparisons can be made? It doesn't take a fancy system—just a cheap spring scale hanging in the barn by a large calendar tacked to the wall. And a pencil chained down so nobody can snitch it.

Write down the pounds or kilograms of milk taken at each milking. Write small, because chances are the calendar will become the place for many day-to-day notes about what is going on with the animals, the land and the weather. Some will be for fun. Much of it may turn out to be very useful information to have at hand.

CHAPTER 14

Milk and the Home Dairy

Milk from a healthy cow is a safe and excellent food for most people, but it has to be handled with care. Being rich in nutrients it's an ideal medium for germ growth. Being delicately flavored and near white it's quick to pick up "off" flavors and to show the smallest specks of hair or dirt that may fall into the pail during milking.

The best milk is kept as clean as possible from the start, is quickly strained and cooled, or else is pasteurized and then cooled.

Begin by keeping the cow clean. Hands should be washed before milking if only in the chlorine solution being used to wash and stimulate the udder.

Equipment

The milk pail should be a milk pail *only,* scrubbed and rinsed—preferably with boiling water—after each milking.

After milking and pouring off what whole milk goes to the cats and calf, the rest should go to the house right away for straining. A couple of thicknesses of any tightly woven, soft cotton cloth will do. Between milkings the strainer also must be thoroughly washed and either sterilized with boiling water or allowed to hang and dry in the sun.

One of the handiest pieces of equipment for any home dairy will be a tall, straight-sided tank able to hold up to about 10 quarts of milk, and fitted with a bottom spout or petcock. Manufactured tanks of this

168

type, called "creamers," came on the market in the 1800s when a "deep setting" method of cream separation took over from the earlier practice of skimming cream off the top of milk put down in cool cellars in wide, shallow bowls or pans.

Somebody may be manufacturing these deep creamers today. If not, they often can be found in farmhouse attics or junk shops. The best have narrow windows above the spouts for reading how much cream has gathered above the "skimmed" milk.

The beauty of any tank with a bottom spout is that milk can be strained into it and then immediately be drawn off into containers for pasteurizing, or into bottles or cans being used to sell raw, whole milk.

Used as a creamer, all of the strained milk can be left in the tank that then goes into the refrigerator, down the well or into a spring-house for fast cooling.

Deep well or spring water may be a little above the recommended 40° F. (4° C.) temperature for keeping milk. However, using water to cool the milk from cow temperature to 50° F. (10° C.) would ease the load on a refrigerator. Don't cover raw milk tightly until it has cooled. This lets "off" odors escape.

Pasteurizing may drive some "off" odors away—and maybe give the milk new odors and flavors of cooking if it isn't done properly. Pasteurization does not kill all the microscopic organisms that may be in fresh milk—only the ones that can infect humans and most of those that cause spoilage. (If you and your milk customers want raw milk but some health ordinance says you can't sell the stuff, how about selling your customers shares in the cow? Then charge for "services rendered" in caring for their animal.)

Gravity separates the cream from the milk in this old-style, "deep-setting" creamer.

Milk Pasteurization

Milk can be pasteurized in three ways. It can be heated for a second to a rolling boil; heated to 170° F. (77° C.) for 15 minutes; or heated to 150° F. (66° C.) for half an hour. The best way, for avoiding cooked flavors, is to heat it over water to 150° F. and to hold that temperature for the half hour. The rolling boil method has to be done over direct heat and definitely will give milk a cooked taste.

Automatic pasteurizers with timers and heat control units take the human effort and guesswork out of pasteurizing but at a high price. A cheap home-pasteurizing unit takes quart canning jars, a regular canning pot and rack.

Fill sterilized jars to within an inch or so of their tops with raw milk. Place them in the canning rack and fill the pot with water to just above the milk line. Suspend a dairy thermometer half-way down inside the center jar. Heat the water until the milk registers 150° or 170° F. (66° to 77° C.) and hold it there for the appropriate time.

When the time is up lift the rack of pasteurized milk into a sink of cool water. Change the water once or twice as it warms, or maybe add ice cubes. As soon as the milk has dropped to about room temperature top the jars with clean caps and store the milk in the refrigerator or down the well.

Getting the Cream

It takes most of the hours between milkings for cream to rise as well as it is going to, sitting quietly in a bottle, creamer or can. Overall the cream will have about 22 percent of butterfat and the milk will have something less than a half of 1 percent.

It only takes five minutes to get light to heavy cream out of milk run through a centrifugal separator. These machines must have been godsends to turn-of-the-century cream farmers. The milk has to be run through the separator while it is warm and before cream has had a chance to form a surface skim.

Separators are fun to play with but they probably aren't worth the work of cleaning for the milk from just one cow. A sink-full of parts

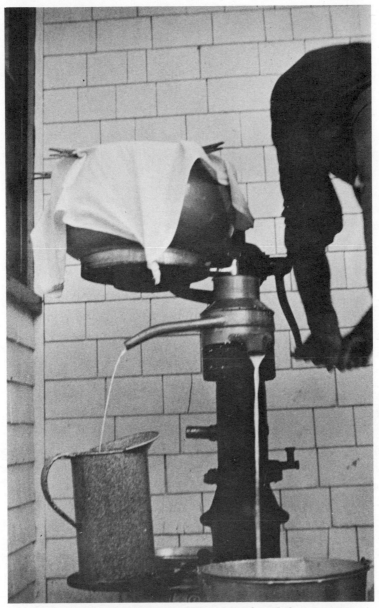

Instant cream! Centrifugal cream separators of the type pictured became widely popular in the early 1900s providing cream producers with a method for removing the butterfat from milk in minutes instead of hours. Several European companies are still manufacturing centrifugal separators for use on small farms.

has to be scrubbed and sterilized at least once a day when the machine is in use. An automatic washing machine might make a difference.

COMPONENTS OF 85 TO 93 SCORE* BUTTER

Fat %	Water %	Salt %	Curd %
82–84	14–16	0–4	0.1–3.5

* In commercial United States butter manufacture the product is rated against this scale from 85 to 93 on the basis of flavor, smell, body, texture, salt, color and packaging. The 85 score butter is a poor product, for one thing having a pronounced, obnoxious flavor.

Making Butter

Cream that has about 30 percent butterfat "makes" into butter faster than heavier or lighter cream. Cream that is heavier than 30 percent fat quickly layers the thrashing surfaces of a churn with a cushion of fat that softens the blows intended to bring fat droplets together. The excess of water in lighter cream acts as a barrier between droplets that have to come together in the butter-making process. There are other factors that speed or slow the butter-making process though, and all things considered, the 22 percent cream off gravity-separated milk is good enough.

Most of us are used to sweet-cream butter. But popular years ago, and still popular in a lot of country homes, was butter made from cream that had just begun to turn sour. It may taste strange at first, but people who get used to this butter often find that the modern, supermarket style tastes flat no matter how much salt is added.

It takes 30 to 40 minutes to make butter with sweet, gravity-separated cream—a little less time with slightly turned cream because the acids produced in the souring process coagulate *casein*, which has the effect of lowering the viscosity of the cream. This allows for greater impact between fat globules.

COUNTRY BUTTER RECIPE

Some butter recipes are so complex they hide the fact that the product is nothing but overdone whipped cream that's been washed and salted to taste. One quart of cream yields up to half a pound of butter.

Fill a churn half-way to three-quarters up with fresh or lightly soured cream that's a little cooler than room temperature. One way or another, agitate the cream until it whips and then breaks, and gathers into particles of butterfat that are pea-size or larger sloshing around in the watery buttermilk. Pour the buttermilk off through a strainer to catch the smaller fat particles.

Gather the butterfat into a large, soft mass using a wooden spoon or clean hands. Put it in a large bowl and pour in cool but not cold water. Work the fat, again with a spoon or by hand, to remove trapped buttermilk. Two or three washings may be needed before the wash water stays relatively clear indicating the last of the milk is gone. Keep working the butter until the water is pressed out to your satisfaction.

Salt to taste. A good way to experiment is with slices of fresh-baked bread. Some people insist on pickling salt. Table salt will do—about a teaspoon to the pound.

Pack the warm butter in shallow pans or whatever and divide with a knife so that after it has cooled in the refrigerator it can easily be snapped into pound or so pieces for wrapping and storing back in the refrigerator or freezer.

BUTTER CHURNS

At least two types of butter churns are on the market today. One is styled after the ancient upright dasher churns. Others, like oversized restaurant milk shakers, come with jars for churning a half to three gallons of cream electrically. Antique, hand-cranked churns can be found, too. In the old days medium-sized barrel and upright dasher churns sometimes were powered by dog or goat treadmills or by Gramp's rocking chair.

A simple churn is a wide-mouthed plastic gallon jar half filled with cream, capped tightly and used for games of toss and catch. The only problem here is that butter-making becomes a daily job, when it doesn't have to be more than an every three- or four-day one.

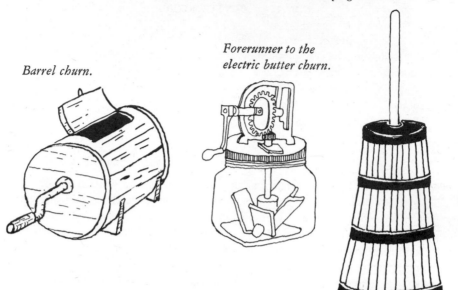

Barrel churn.

*Forerunner to the
electric butter churn.*

Upright dasher churn.

FACTORS THAT SPEED OR SLOW
THE BUTTER-MAKING PROCESS

BUTTER CHURNS FASTER IF:

Cream is from the cow early in lactation, therefore has larger fat globules.

The cow is on lush pasture or succulents so that she produces cream with more large globules that are soft.

The cream has been allowed to "ripen" to the point that it is about 0.4 percent acid.

BUTTER CHURNS SLOWER IF:

The cream is cold. It should be only a bit below room temperature.

Cream is under 20 or over 40 percent butterfat.

Cream is fresh and "sweet."

Cow is late in lactation, yielding a cream with small fat droplets.

Cow is on grass hay or other "hard" feeds that may produce harder fat globules. Cottonseed meal and coconut oil meal are said to produce a hard butter if fed in large quantities.

STORING BUTTER

Butter can be stored for months in a freezer. If it is heavily salted, or lightly salted and packed in a salt and water solution that floats a medium-sized raw potato, it will keep for several months in a cool cellar. Sixteenth-century Icelanders kept salted butter through the winter buried underground in wooden caskets up to 30 feet long.

A nineteenth-century English farmer suggested that butter for storage be mixed with pickling salt at a rate of 10 ounces for every 14 pounds of butter (more or less, depending on how long the butter is to be kept). But that's going to be cooking butter unless you are a salt fiend.

Sour Milk

Raw milk or cream should keep fresh easily for four or five days if it is cooled quickly and kept in a good refrigerator. If milk or cream sours sooner than seems reasonable, check to make sure pots and pails are being kept clean. Once soured milk gets moving around a kitchen it may take an unusually thorough and simultaneous scrubbing and sterilization of all the milk-processing utensils to stop the invasion.

Another cause for rapidly souring milk can be an infection in one or more quarters of the cow's udder. The general term for these infections is *mastitis;* see Chapter 17.

So many kinds of microorganisms like to live in milk, whether it is raw or has been through pasteurization, it's just fortunate most of them are not harmful to humans. Better still, once "good" organisms (certain harmless bacteria and some yeasts) get a start in milk they can wipe out disease germs and even turn milk into foods that are easier to digest.

As milk turns, it picks up different flavors, depending on what organisms or enzymes have gained the upper hand. Long before the processes were understood, people learned ways to control the turning so that they could produce good, safe cultures.

One of the simplest methods was to "start" milk by seeding it with a spoonful of the best-tasting culture from the week before. A popular culture spooned down through the centuries is yogurt (see recipe).

Although yogurt has a few more days of food-for-people useful-ness than milk, it does spoil rapidly. In part this is because yogurt is mostly water. And so, like milk, it is a fine home for the progression of living systems that eventually return all organic matter to the soil. Maybe this is one reason why cream cheese was invented. Essentially it is a drip-dried yogurt.

YOGURT RECIPE

Skimmed milk to light cream can be used for yogurt. The "starter" can be commercial yogurt starter or a tablespoon of commercial plain yogurt for every pint of yogurt you plan to make. Thin the "starter" in a cup of milk.

Warm the light cream or milk to 95° to 110° F. (38° to 43° C.) in a glazed pottery, glass, tin or enamel bowl or pot. Stir in the "starter."

One way or another let the mixture sit at its ideal 95° to 110° F. temperature for several hours until the yogurt bacteria have worked their thickening, flavoring wonders. Overnight in a wood stove's warming oven is Susan's favorite. Some people line the bottom of an insulated picnic basket with wrapped hot bricks and place the developing yogurt in there. Others use a double boiler arrangement with the water over low heat.

The usual ripening period is 6 to 10 hours. A shorter period will generally produce a softer, milder-flavored yogurt.

CREAM CHEESE RECIPE

Depending on the flavor wanted in the final product the "starter" for cream cheese may be a cup of yogurt, commercial buttermilk or something made up from a commercial cheese starter. Mix with 2 cups of warm milk and let stand 24 hours.

Add two quarts of warm milk and let clabber another 24 hours. Warm over hot water for half an hour and pour into a cheesecloth bag to drain. Let stand for one hour. Salt to taste and wrap in waxed paper. Refrigerate. (From Phyllis Hobson's *Making Homemade Cheeses & Butter,* Garden Way Publishing, 1973.)

ENGLISH CREAM CHEESE

Pour one quart of cream in the top of a double boiler. Add hot water to the bottom pan. Heat the cream slowly almost to the boiling point.

Remove from heat and add one rennet tablet dissolved in cold milk. Stir well and let stand until thick. Then break the curd slightly with a spoon and pour into a drain bag. Let drain for 24 hours, then press in a cheese press under light weight for another 24 hours.

Remove from press, wrap in cheesecloth and rub flake salt over the cloth. Hang to dry one or two days before slicing.

As any milk product ferments or loses whey it gains friends among the people doctors say are "lactose intolerant." Usually these people are adults who apparently do not produce enough of the enzyme *lactase* that's needed to break down the milk sugars in fresh whole or skimmed milk. They may experience anything from mild gas to cramps if they drink too much milk all at once.

When milk curdles (clabbers), almost all of the lactose goes with the albumin and globulin proteins and B vitamins into the liquid whey. Left behind are the firm, easily digested curds of casein protein rich in minerals and having most of the butterfat that was in the original milk.

Cottage cheese is probably the simplest of all curd cheeses to make. The soft curds may be formed by allowing naturally-occurring bacteria to turn some of the lactose into lactic acid (any acid will curdle milk)

DOT'S COTTAGE CHEESE RECIPE

Take some of the evening's milking and let it stand in the kitchen overnight. Add about an equal amount of fresh morning's milk and leave the whole standing at room temperature until it is well clabbered —perhaps another 8 or 12 hours.

Break up the curds with a knife or wisk and pour in a couple of cups of boiling water while stirring with a wooden spoon. Test the curds for firmness with clean fingers. If they're too soft, pour in some more boiling water. When the curds seem right, strain through cheesecloth or strainer, wash and salt to taste.

COTTAGE CHEESE RECIPE
(about one quart)

Heat a gallon of skimmed milk to 75° to 80° F. (24° to 27° C.) and add a half or a whole rennet tablet dissolved in water or about a half cup of yogurt thinned with warm milk. Let cool.

Let the mixture stand at room temperature through the day or overnight. It should form a soft curd in this time. Cut the curd with a gentle slicing action of a wire wisk or use a long knife to cut repeatedly through the curd up, back, down, and sideways, creating something on the order of a pot full of inch cubes floating in whey.

Slowly heat the curds and whey, stirring a bit (and gently) to prevent burning. When the temperature gets up to about 100° F. (38° C.) the curds will begin to firm up. Keep the curds moving now until they are firm enough not to break apart easily when they are pressed together.

Careful! Too much heating at this point makes squeaky-hard curds. When they seem firm enough but not tough, separate the curds and whey by pouring the two through cheesecloth or a strainer.

Flush the curds with cool water and let them drain. When they have stopped dripping add salt to taste and maybe some cream to moisten the cheese. Done.

—or faster (and usually necessarily with pasteurized milk) by the addition of commercially prepared rennin.

RENNET

Curdling by adding rennin may have been discovered by the first person who tried to carry a batch of milk home in a calf's stomach, an obvious choice of container before the inventions of pottery or glass. Today most commercial rennin, sometimes called *rennet,* is processed from the lining of calves' stomachs.

Years ago it was done on the farm. When a calf died or was slaughtered, the stomach (abomasum) was removed, cleaned, and was either salted and dried or was stored in brine. A small slice of dried or pickled stomach rinsed and then steeped in hot water produced a brew that was loaded with rennin for starting cheeses.

HARD CHEESES

Of all milk products, hard cheeses are the least perishable and easiest to transport because they are dry. They are the easiest to digest because the lactose has been removed or has been converted to lactic acid in the ripening process. They are the safest of unpasteurized products too, because their dry, acid environments are least suited to the needs of streptococcus, tuberculosis and brucellosis bacteria.

Most of the hard cheeses go by the names of the farming or cheese manufacturing centers where they were invented, but shown on the next page is a sort of mongrel hard cheese based on a recipe provided by the Vermont Extension Service.

A simple 1-pound cheese press using a brick or old iron for weight.

THE WHEY

Although we usually feed whey to chickens or pigs (and it is an excellent food for these animals), centuries ago it was a popular drink on many country farms. Here is a recipe for whey lemonade taken from Phyllis Hobson's *Making Homemade Cheeses and Butter*.

WHEY LEMONADE

Strain 1 quart of whey. Add the juice of 2 lemons and sweeten with about 6 tablespoons of honey or sugar. Chill.

HARD CHEESE RECIPE

Heat eight quarts of whole milk in an enameled or tin pot or pail to 85° F. (30° C.). Add a quarter of a rennet tablet (or similar renneting agent) dissolved in a half cup of cold water. Mix thoroughly. Set the pot or pail in warm water (85° to 90° F.) and let stand until a firm curd forms—about 30 minutes. To test the firmness put your finger into the curd at an angle and lift. If the curd breaks cleanly over the finger it is ready to cut.

Cut into squares of about three-eighths inch with a long knife. After cutting both ways vertically, use the knife at an angle following the earlier slice marks in both directions in an attempt more or less to cube the curd from top to bottom.

Heat the water in the outside container slowly, taking about a half hour to raise the temperature of the curd from 90° to 100° F. (36° C.). Stir the curd with your hand very gently from time to time to keep the temperature even throughout and to keep the pieces of curd from sticking together.

When the curds are firm enough so that they will not stick together, pour them into a cloth to drain off the whey. Form the curds into a ball inside the cloth bag and hang for two or three hours while the remaining whey drips out.

Take the ball out of the bag and place it on three or four thicknesses of moistened cheesecloth. Take another piece of cheesecloth and fold it to make a long "bandage" about three inches wide. Wrap this tightly around the ball of curd and pin in place. Fold down at the top and crumble the top of the curd ball until it is perfectly smooth and there are no cracks extending into the center.

Lay a wet piece of folded cheesecloth over the top of the ball. Place a flat plate on top and weight down with the equivalent of a brick or flat iron. (You may find that the weight tends to slip to one side giving the cheese an odd shape. If so, make a simple cheese press from two boards and a round stick.) Your round loaf of cheese should not be more than six inches across or it will dry out too much. That night turn the cheese and place the weight on top again. Let it stand until morning.

Remove the cloth bandage and place the cheese on a board in a cool, frost-free place like the cellar. Turn once or twice a day until a rind is formed. This probably will take three days. Then rub a tablespoon of salt into the cheese two days in succession. After this, rub thoroughly with a small amount of butter. Do it again the next day. Rub and turn the cheese each day until the rind is very firm. After a week or two it will not be necessary to rub so often. Two or three times a week will keep it from getting dry and prevent mold from developing.

CALF'S MILK CUSTARD

Last, though first in the milking season, comes "calf's milk custard" made from colostrum. The basic recipe comes from Dale Eriksson of Greenfield, Mass., whose grandmother brought it from Sweden. The recipe calls for colostrum from the second, third or fourth milking. "The darker the color the richer the colostrum."

CALF'S MILK CUSTARD

Mix one cup of regular milk with two to three cups of colostrum. Add about ½ cup of honey or sugar and stir until dissolved. Pour into a double boiler and sprinkle a little cinnamon on top. Cook without stirring until thick—about 2 to 3 hours.

The custard works without eggs or stirring because the albumins in colostrum thicken on heating. Since the albumin content of colostrum drops fast through the first days of milking a lot of combinations of colostrum to whole milk and cooking time are possible.

Calf's milk custard can be baked, too, following the directions in any cookbook for baking egg custards. Try their suggestions for seasoning egg custards, too—like using nutmeg instead of cinnamon, or with the mixture poured over sliced, cooked or canned peaches. (Unfortunately many commercial farmers have taken up the practice of routinely shooting their cows' udders full of antibiotics at the time of drying off. People who either react to antibiotics or just don't want them in their diets for any reason shouldn't use colostrum from these cows.)

CHAPTER 15

Breeding and Heat Periods

A heifer comes into heat for the first time at around 10 months of age. From then on, until she is successfully bred, she will come into heat about every three weeks (17 to 24 days).

Because the heifer still has a lot of growing to do even after the age of puberty has arrived, attempts at breeding usually are held off until she is 15 months or older. Many breeders go by the weights of their heifers. The small breeds may be bred when they reach 600 to 650 pounds. Let heifers of the intermediate-sized breeds go to 700 pounds, and those of the large breeds to 750 to 800 pounds.

A cow just calved may come into heat any time in the next month. However, to make sure that her system has had a chance to recover from the last gestation don't attempt to breed her back for at least 60 days after calving. You could let her go 90 or 100 days, thereby assuring yourself and the cow even a longer rest and dry period at the end of the lactation period just started.

Each period of heat will last one to one-and-a-half days. The best time for conception lies within this period, beginning six to eight hours after the start of the period and continuing about 10 hours for natural breeding and up to 18 hours for artificial breeding. In general young cows have shorter but more intense heat periods than older animals.

There are many, many ways people have for telling when their cows are in heat. Some say they can see it in their cows' eyes. More often people look for an agitated cow who swishes her tail and bellows repeatedly for no other good reason. The cow's vulva may become swollen and blushed.

NORMAL GROWTH IN WEIGHT AND
HEIGHT OF FEMALE DAIRY CATTLE

	AYRSHIRE		GUERNSEY		HOLSTEIN		JERSEY	
Age	*Weight (lb.)*	*Height (in.)*	*Weight (lb.)*	*Height (in.)*	*Weight (lb.)*	*Height (in.)*	*Weight (lb.)*	*Height (in.)*
1 mo	89	28.6	77	28.2	112	30.6	67	27.0
2 mos	119	30.2	102	29.8	148	32.3	90	28.9
4 mos	198	34.0	173	33.5	243	36.2	158	32.6
6 mos	293	37.2	260	36.9	355	39.7	243	36.2
8 mos	389	39.9	350	39.9	462	42.3	324	39.0
10 mos	469	41.7	427	41.7	552	44.4	393	40.9
12 mos	538	43.2	490	43.3	632	46.0	450	42.2
18 mos	725	46.5	663	46.4	845	49.3	601	45.2
24 mos	902	48.3	818	48.0	1069	51.7	733	46.9
3 yrs	968	48.7	901	49.9	1165	53.0	855	48.2
4 yrs	1035	50.2	990	50.4	1232	53.3	897	48.5
5 yrs	1080	50.4	1055	50.6	1330	53.6	937	49.0
6 yrs	1132	49.1	1093	49.7	1317	53.7	973	48.4

(From Agriculture Canada Bulletin 1349.)

A cow in the midst of a strong heat may hold back on her milk almost entirely for one milking. There may be a clear mucous discharge from her vulva that is slippery to the touch—rather than being sticky like the kind of discharge that sometimes occurs during gestation.

Standing Heat

A few hours after the earliest signs of agitation and bellowing may have been noticed a cow normally will go into what is called her "standing heat," which may last 12 hours or more. During this time she will mount other cows and later will stand still when they mount her. At any other time a cow that is mounted will scoot away.

Natural breeding works best after the cow has been in standing heat for three or four hours and on, until she will no longer stand still for the bull. A best rate of success with artificial insemination begins at the same time but continues as much as 6 to 12 hours after the cow is no longer "standing."

Standing heat is what all owners of several cows look for. In some cows it may be the only outward sign noticeable enough when you're working with dozens of animals. Hopefully the owner of one cow will be able to pick up on more subtle changes in the cow's looks or behavior. But even with one cow you can watch for a standing heat. Or maybe I should say watch out for it, because if she's in a good strong heat she may try to mount you. The first time a cow almost got me I couldn't imagine what had come over her. There'd been no other signs of heat. Fortunately there was a stanchion in the way.

If a cow is thought to be in heat, stand alongside holding her collar, and scratch her brisket. With any luck she'll lean back in preparation for a lunge. Another sign of heat the family cow owner can look for is the cow's jumping the fence and joining a neighboring herd.

A day or two after a heat period, bred or not, a cow may show a bloody discharge. It may be slight and only barely noticeable, dried on her flank or tail. It may be the only sign that the cow was in heat. But that's good. It's a start. Because then you can look for the next heat within 16 to 22 days.

Breeding and Gestation Times

Considering all that's been said, when it comes to having the cow bred figure it this way: If you see the cow is in heat in the early morning, she'll want to be bred late in the day. Call the artificial inseminator right away or make arrangements to take the cow to a neighborhood bull. If the cow is noticed coming into heat in the late morning, that night will be good for natural breeding. The following morning will be good for AI. If she's noticed in the evening, shoot for natural breeding that night and any time the following morning for AI.

If three weeks after breeding no signs of heat or bloody discharge are noticed it's likely all is going according to plan. Seven to nine weeks after breeding a vet will be able to detect a developing fetus if you want the assurance. And after about five months' gestation it will probably be possible to feel or see the calf punching around. It always seems I get the message first through my head as I lean into the cow during milking.

Breeding Problems

Some cows are difficult breeders. First, they may be the kind that throw what are called silent heats that are very difficult to detect. Others throw wonderfully vigorous heats but for one reason or another fail to "settle," which is another word for having conceived. I took one heifer through five heats and breedings, three AI and two natural, before a bull finally did settle the matter.

A few of the more common reasons why cows won't conceive are:

1. *A heifer is a freemartin.* These are heifers born twins to bulls. About 90 percent of these heifers are infertile and have incomplete reproductive systems due to interference from male hormones that somehow were able to pass from the bloodstreams of fetal bull calves into the bloodstreams of their sisters. Derivation of the term *freemartin* is not certain. Some think *free* may be a corruption of *farrow,* meaning barren, and that *martin* may have to do with the fact that years ago in England St. Martin's Day (November 11th) was a big day for slaughtering cattle for winter food. Freemartins are supposed to produce an excellent, fine-grained beef.

2. *Persistent "yellow body,"* or *corpus luteum* in the ovary. This little growth that develops in the pocket of an erupted follicle that ejected the last egg, produces hormones that keep any more eggs from being shed. It's supposed to persist through most of the gestation period. But after a heat, if a cow is not bred, the yellow body is supposed to disappear. If it doesn't dissolve, the ovaries may remain dormant. A veterinarian's diagnosis and treatment will be necessary to find and correct this problem.

3. *Ovarian cyst.* Actually this is an egg in its ovarian pocket that for some reason, though ready to be released, does not break free to pass on into the uterus. Because it is ready, hormones from the pocket throw the cow into a heat that goes on and on. These "cysts" can be removed by a vet.

GESTATION TABLE FOR COWS

Breeding dates	Jan.	Feb.	Mar.	Apr.	May
Day 1 _____	Oct. 10	Nov. 10	Dec. 8	Jan. 8	Feb. 7
2 _____	11	11	9	9	8
3 _____	12	12	10	10	9
4 _____	13	13	11	11	10
5 _____	14	14	12	12	11
6 _____	15	15	13	13	12
7 _____	16	16	14	14	13
8 _____	17	17	15	15	14
9 _____	18	18	16	16	15
10 _____	19	19	17	17	16
11 _____	20	20	18	18	17
12 _____	21	21	19	19	18
13 _____	22	22	20	20	19
14 _____	23	23	21	21	20
15 _____	24	24	22	22	21
16 _____	25	25	23	23	22
17 _____	26	26	24	24	23
18 _____	27	27	25	25	24
19 _____	28	28	26	26	25
20 _____	29	29	27	27	26
21 _____	30	30	28	28	27
22 _____	31	Dec. 1	29	29	28
23 _____	Nov. 1	2	30	30	Mar. 1
24 _____	2	3	31	31	2
25 _____	3	4	Jan. 1	Feb. 1	3
26 _____	4	5	2	2	4
27 _____	5	6	3	3	5
28 _____	6	7	4	4	6
29 _____	7	—	5	5	7
30 _____	8	—	6	6	8
31 _____	9	—	7	—	9

(From *Agriculture Canada,* Bulletin #1439 of 1971.)

When a cow is bred, write the name of the month in the appropriate space in the first column. Using a straight-edge, follow across the table to the col-

June	July	Aug.	Sept.	Oct.	Nov.	Dec.
Mar. 10	Apr. 9	May 10	June 10	July 10	Aug. 10	Sept. 9
11	10	11	11	11	11	10
12	11	12	12	12	12	11
13	12	13	13	13	13	12
14	13	14	14	14	14	13
15	14	15	15	15	15	14
16	15	16	16	16	16	15
17	16	17	17	17	17	16
18	17	18	18	18	18	17
19	18	19	19	19	19	18
20	19	20	20	20	20	19
21	20	21	21	21	21	20
22	21	22	22	22	22	21
23	22	23	23	23	23	22
24	23	24	24	24	24	23
25	24	25	25	25	25	24
26	25	26	26	26	26	25
27	26	27	27	27	27	26
28	27	28	28	28	28	27
29	28	29	29	29	29	28
30	29	30	30	30	30	29
31	30	31	July 1	31	31	30
Apr. 1	May 1	June 1	2	Aug. 1	Sept. 1	Oct. 1
2	2	2	3	2	2	2
3	3	3	4	3	3	3
4	4	4	5	4	4	4
5	5	5	6	5	5	5
6	6	6	7	6	6	6
7	7	7	8	7	7	7
8	8	8	9	8	8	8
—	9	9	—	9	—	9

umn of dates that falls beneath the month she was bred. For example: A cow successfully bred September 12 may be expected to calve on or about June 21.

4. *Vitamin or mineral deficiency.* Serious deficiencies of vitamin A or of phosphorus are the most frequent causes of infertility that can be blamed on nutrition. Most likely they would be found in cattle being kept primarily on poor-quality roughage, be it straw, weathered hay or weathered brown pastures.

Artificial Breeding

The technique of artificial breeding in cattle came into wide use in North America only after World War II, and has made possible incredible changes in breeding. The semen from one good bull now can be used to breed thousands of cows a year. Through natural breeding a bull might be good for no more than 50 cows a year.

AI has its potentials for bad as well as good. A bad trait, at first unnoticed, might be bred into thousands of offspring before being discovered. Cloning and uterine implantation systems, still in the experimental stage, will require even more care to avoid the rapid spread of genetic horrors like the hip displasia that has plagued some pure breeds of dogs in recent years.

With proper controls AI is a beautiful thing for the owners of family cows who for ten dollars or so can have a choice of quality bulls. The breed of bull to use depends on what is wanted in the calf and on the capabilities of the cow. Don't breed a small cow to any breed of bull likely to give her a large calf and therefore a potentially difficult birth. For instance, I wouldn't breed a Jersey cow to a Holstein or Brown Swiss bull. Some Holstein farmers regularly breed their heifers to Angus bulls to assure small calves the first time around.

When we have wanted beef we have bred our cows and heifers to Hereford or Angus bulls. A calf of either sex from a dairy-beef cross likely will yield more and fatter meat than will a straight dairy animal. However, if in breeding for a pure dairy animal you get a bull calf, don't hesitate to raise him for the very good and cheaper beef than could come from any store.

CHAPTER 16

Calving and Calf Feed—
Care of Young Stock

In spite of eight thousand years of domestication, cows still know how to have calves. Interfere with calving as little as possible.

Nine times out of ten a cow that's been well fed and had a six-week rest from milking will have her calf quietly in the barn some night, or in a secluded corner of the pasture. By the time she's discovered, the calf will be dried off and looking for a meal.

There usually are ways to tell when the cow is about to freshen. First of all it should happen within nine-and-a-half months of breeding, give or take a week. If it isn't known just when she was bred, keep an eye on her udder. The cow should begin to "spring" or "bag up" within three or four weeks of calving. The udder swells, and a day or two before calving it will look uncomfortably swollen and rigid. The teats may stand out and in some cases milk starts to dribble.

A day or so before calving there will be a noticeable falling away between the hips and tail bone. Where there was a ridge there is a hollow you can drop your hand into. The calf is moving back and the pelvis is parting to make way for the birth.

Other last-minute signs are: The cow becomes fidgety and may not be interested in eating; the lips of the vulva swell and there may be a discharge of mucous; the cow's temperature may drop a degree or two below normal.

These are some of the signs. But don't count on them. I had a heifer that "bagged up" two months before she calved. Not great, swollen and dribbly, but certainly looking imminent. Then there's the cow that waits until the last minute. In the morning she looks a little bagged up but mostly like her old self, and that afternoon she's got a calf.

If the weather is warm and the ground dry I like to let a cow have her calf in the pasture. (This is assuming the pasture is small and near the barn.) We keep an eye on her when she's due, checking on her now and then through the day, late at night and first thing in the morning.

This is taking a chance. It could happen that the cow could go into labor, have difficulty and go down where it's least convenient to help her out.

The best place for a cow about to calve is in a well-bedded box stall about 10 feet square. A lousy place is in a stanchion. If the cow calves during the night she can't get at it to dry it off and nuzzle it to life. I've seen calves that had drowned in the gutter. If there is no other place for the cow to be, fill the gutter with bedding or hay.

In a normal and unassisted birth the cow may stand throughout. She begins to strain a half hour or so before the encasing "water bag" and front toes of the calf appear, pads down. The calf emerges little by little—it's like a coming tide. There's the toes . . . oops. Gone again. Then comes the nose, snuggled above and between the front toes.

In another half hour to an hour from the time the front toes appear the calf will be born. The time isn't important as long as there is steady-by-jerk progress.

Helping the Birth

If progress ceases—nothing has changed for an hour, say, and/or the cow appears to have given up—then it's time to step in to help. Also the cow will need help if the calf is coming out the wrong way, which would be indicated as soon as a hoof appeared pads up, or if there weren't that three-point appearance of muzzle and front hooves. (One possible exception to this always "correct" presentation may be in the case of twins. Here the second calf may appear hind-feet-first, toes up, and without a hitch. You may expect twins if the calf just born is on the small side—though this will be hard for the inexperienced to judge.)

It may be a simple matter to help the cow that has given up on an otherwise normal birth, especially if the front hooves are well exposed.

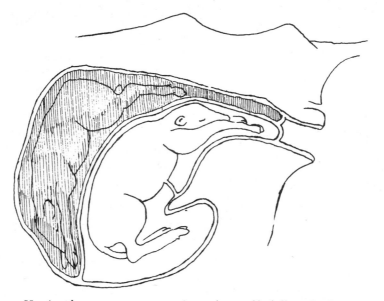

Here's the correct presentation of a calf followed (in shadow) by a twin that may as often come hind-end first as the normally "correct" way. The reason for there being more flexibility with twins is because the calves are almost always far smaller than those born singly.

Place slip knots of rope or stout twine over each front foot and pull back and down when the cow strains. It will take some force, but work up to what is needed rather than giving all you have in one great jerk. The less outside hauling needed the better for the cow and calf. Even if the cow has stopped straining, space your own efforts around intervals of rest. The calf's head and shoulders have the greatest difficulty passing through the pelvis of a small cow. Broad shoulders will pass more easily if more pull is applied to one front leg than the other.

When there is a poor presentation an amateur might soap up an arm and explore within the cow to see what's wrong. But it can be a very strenuous and tricky task straightening out a calf. If you must go in because there is no more experienced person to help, go in between the cow's strainings and only attempt to shift limbs or the body of the calf when the cow is relaxed.

Helping the Calf

A lot of books would have you rush in with scissors and sutures to operate on the umbilical cord as soon as the calf is born. Forget it. I've never seen one that didn't snap of its own accord just the way nature intended, eight inches or so from the calf's belly. It will bleed out some. Let it. Normally it won't bleed for very long. The natural stretching and snapping causes the broken blood vessels to contract and brings on rapid clotting.

If for some freak reason the cord doesn't break, or it does and the bleeding is a flood that doesn't appear to be lessening in a matter of minutes, then do tie the cord off near the calf's belly. After tying, cut the cord a few inches down. That longer, naturally snapped cord that may be dangling almost to the floor is no problem. It dries and shrivels up in a matter of hours and will drop off at the calf's belly within a week or ten days.

If you are there when the calf is born, or as you find it, it is a good idea to dip the cord in tincture of iodine to prevent infection. Put the iodine in a juice glass or the like and bring the glass up under the cord, right up to the calf's belly so that the whole cord is immersed.

The next warning often given is not to let the cow eat the afterbirth, or placenta (also known as "cleanings") which she will expel after the calf is out. It is a natural instinct for animals to eat the placenta. It's thought to be a sort of cleaning-up exercise that keeps the wolves away.

I have been told that it is possible for the placenta to plug the cow's gastrointestinal tract, but I have never interfered with the birth process this way. Something is telling them to do it. We can guess why, but we don't know all there is to know about cow physiology. If the placenta were plastered with filth I'd take it away. But there shouldn't be any panic about being there when the placenta arrives.

When the calf is born the mother's first attention is toward cleaning and drying it off. A person on hand can step in when the calf drops to wipe off its nose and to assist with a little artificial respiration if the calf is not breathing. Either mouth-to-mouth or the old-fashioned push and relax on the rib cage assistance could be used. If the weather is cold and nasty it does help to rub the calf down with some rags. Burlap bags are great for this, but they are becoming scarce.

The Calf's First Feedings

Within an hour or so the calf will be up and looking for its first feed of colostrum, the first and vitally important milk that's rich in vitamins and in antibodies that the calf, unlike a human baby, was unable to absorb from the mother's circulatory system during gestation. The antibodies are needed to protect the calf from the most common germs that are everywhere in its new world.

Give the cow's udder a washing if it is needed, and maybe help guide the calf to a teat if it's having a rough time of it. Certainly people watching this first groping search will be more pained than the little calf.

It is essential for the health of the calf that it get a feed of colostrum within 24 hours of birth—the sooner the better. After 24 hours a change comes about in the calf's intestines and they can no longer absorb the needed antibodies.

On large farms the calf is often taken from the mother immediately, even before it gets its own taste of colostrum. With the family cow and calf things can be and usually are more relaxed. The cow and calf may be separated after the calf has had some colostrum and be fed from a bucket thereafter or the two may be left together for three or four days or more before bucket feeding begins.

A third alternative is to take the calf away after it has had some colostrum and then to bring it back to nurse after regular milking times. Two teats are milked into a bucket and two are left for the calf. The calf is given a half hour or so with its mother before being taken back to its corner.

If the milk comes on strong and there is more than enough for the calf in two teats, milk three into the bucket, or adjust whatever way is necessary to leave the calf about the amount of milk that is called for in the feeding table.

LEAVING THE CALF WITH ITS MOTHER

There are two reasons for taking a calf from its mother—to save milk and to make milking and calf feeding simpler. She doesn't have to be taken away from her mother at all. This past summer a man down the road was keeping a cow and calf together in a pasture. For the time being he had no reasonable way to keep them apart. He would milk his cow morning and night while the calf looked on, fat from her own feeding whenever the spirit moved. The last I heard the man was getting about 14 quarts a day above what the calf was taking.

Since bucket feeding schedules strictly limit a calf's intake of milk, I wondered how it could be that this calf wasn't being destroyed through eating more than she should of that laxative food. A friend explained it by comparing milk to Ex-lax. "If you took a pinch of Ex-lax now and then through the day, no problem. But you sit down and eat a whole bar of Ex-lax all at once, well watch out!" Given the opportunity to eat all she wants, a calf takes a sip now and then, and through the day drinks far more than she could handle in two or three gross feedings. She takes what she needs and she leaves the rest. And if that isn't taken by hand milking or another calf or two, the cow slows down her production.

One exception to letting a calf have all the whole milk she wants, either as a young thing nursing or an older calf on a bucket, may be with female Jersey calves. Some of the Jersey literature warns against overfeeding their heifer calves because, they say, there can be a buildup of fat in and around the rudimentary udder which could interfere in later life with the development of milk-secreting tissues. They also warn that overfeeding may create a fat, heavy udder that overtaxes ligaments holding the udder against the mature cow's body.

I would think this might be more of a concern for people raising registered Jersey heifers for possible sale. It may be that only the Jersey people worry about fat calves because their cows produce such rich milk.

If there is no need to take the calf from the cow, it seems more reasonable to leave them together at least as long as the mother is producing colostrum. This is certainly easier and better if the cow is in a box stall. If the cow is in a tie or stanchion stall it seems more reasonable to let the calf do her own nursing at milking time these first few days than to feed her from a bucket. I think it is better both for the calf and for the sake of the mother's swollen udder; this way the udder gets a natural massaging from the calf.

It is interesting to watch a newborn calf and its mother and wonder if the act of birth has really made them separate individuals. The fact is that, given the opportunity for a natural order of things, the calf is still very much tied to its mother both for the getting of nutrients and for the ridding of solid, or what might be called non-metabolic wastes.

For the first few days the calf always lines up front to rear with its mother when it nurses. Immediately the cow leaves whatever else she may be doing and begins to lick the calf's anus. It may strike people as awful queer to sit around dwelling on a thing like this, but it is such a pattern, seen in dogs and cats too, that there must be a purpose. As the cow licks the calf's tail bobs in the air and its bowels begin to move. The cow cleans the calf thoroughly. It may be, as in the instinct to eat the placenta, that the cow is doing something that in the wild served to protect the calf from predators. But there could be more to it than that. The mother's licking may be a helpful stimulation that triggers hormones that throw the peristaltic action of the calf's large intestines into gear. Certainly the cleaner calf is less likely to get into trouble with flies or infectious microorganisms looking for a home beneath a matted tail.

Bucket Feeding

There are arguments about whether it isn't easier to teach a calf to drink from a bucket if it hasn't learned to nurse. I've worked with calves that had nursed a week and with some that had never nursed at all and I think there simply are some calves that catch on to bucket feeding right away and others that are too foolish and anxious to be of any help to themselves.

In teaching the calf to drink from a bucket, it's first important to have warm milk (your own skin temperature), and preferably straight from the cow. Back the calf into a corner and straddle it with your legs to keep it in there. Hold the bucket in one hand and get the calf

Nipple pail for calf feeding.

sucking on the fingers of your other, palm cupped over its nose. Now raise up the bucket and bear down with your palm.

The calf will dive and dodge and kick. It will butt in frustration, just as it does when it wants its mother to let down her milk. And likely the first attempt will get more milk on the teacher than into the calf. Although the urge may be there, try not to drown the calf. Lower your fingers into the milk just to the point where hopefully some will be drawn into the calf's mouth. As the calf sucks, pull your fingers away.

If the calf doesn't catch on that first time, a few hours more on an empty stomach will improve the situation for the next round. It won't hurt the calf at all. Some calves need four or five lessons before they go to the bucket without the guiding fingers.

The milk bucket must be scrubbed between feedings and should always be held for the drinking calf, or placed in a sturdy bracket a

DAILY CALF FEEDING SCHEDULE

	MILK OR MILK REPLACER (Mix by Directions)				
Age	Calves over 90 lbs.	Calves 70–90 lbs.	Calves under 70 lbs.	Calf "starter"*	Hay
Birth to 3 days	with Dam			0	0
3 days to 3 weeks	9 lbs.	8 lbs.	6 lbs.	All it will eat to 3 lbs. daily. Feed in grain box.	All it will eat. A fine-stemmed, leafy, green mixed hay fed in a hay rack is ideal.
3 weeks to 4 weeks	7 lbs.	6 lbs.	5 lbs.		
4 weeks to 5 weeks	5 lbs.	4 lbs.	4 lbs.		
5 weeks to 6 weeks	4 lbs.	3 lbs.	3 lbs.		
6 weeks to 8 weeks	2 lbs.	2 lbs.	2 lbs.		
2 months	Optional feeding of milk or milk products when available			At 2 mos. increase gradually to 5 lbs. daily.	

* Or any dairy concentrate that does not have urea included as a source of protein.

(Adapted from University of Vermont Extension Bulletin.)

foot or so off the floor. If it's not held, the calf will tip the bucket over and beat it to death.

Rather than teach a calf how to suck milk out of a bucket some people use a nipple pail, a special pail with a long rubber nipple sticking out from the side near the bottom. The warnings against them are that they are hard to keep clean and that older calves can haul the nipple out of its socket.

Check Table 13 for a whole milk feeding schedule and also Chapter 17 for notes on calf feeding and scours. Some people want to switch the calf to skimmed milk or to a commercial milk replacer, and this is easily done after the calf has had a month to six weeks on whole milk. The switch should be made gradually over a couple of days by cutting the whole milk with whatever is to replace it.

Calves two months or older can be accustomed to cool milk. They can be fed buttermilk or even sour milk but only if there is a constant supply of these foods, because it isn't good to switch their systems back and forth. Since vitamin A and some little D value is in the fat portion of milk, the calf going on skimmed should get an A & D supplement unless it's eating two or more pounds daily of green and leafy sun-cured hay.

Weaning to Hay and Grains

After the first week the calf should have a small manger of the best hay. If the regular hay being used for the cow is not good quality, it would be smart to buy a half dozen bales or so of the fine, leafy hay, preferably with legumes in it.

The calf also can be introduced to grains after a couple of weeks. Put a little box with a touch of concentrates in her manger. Put some on her nose so she learns what it's about. Or put a taste in the bottom of her milk bucket after she's lunched. There are special calf starter mixtures but this seems unnecessary if the calf is on an ample whole or skimmed milk diet. Some of the same concentrates that the cow is getting will work perfectly well so long as the concentrates do not contain urea, a fact that should be noted on the feed tag.

Until the calf learns to eat solids, clean the feed box and manger out daily so that it doesn't fill with moldy and unappetizing foods.

Water Needs

From the time the calf begins eating solids it should also have a bucket of fresh water, also fastened off the floor where it won't get dirty or tipped. The calf won't drink much at first, but still the water should be there and should be changed regularly for when she *is* ready. Whenever a bucket with a bail wire is left with a calf the bail should be flipped away from the calf. Otherwise in fooling around she may get trapped with the bail over her head and her face stuck inside the bucket.

Weaning can come any time after a couple of months, though it won't hurt to give the calf some milk at least once a day right on through five or six months of age. At the time of weaning the calf should be strong and living happily on hay. It is better for the calf to be getting a pound or two daily of grains or concentrates as well—more if the calf wants, though I think it is better to follow the same idea as with feeding a cow, which is to feed the concentrates twice daily (and never more than they are cleaning up in about half an hour). The weaning can be done abruptly but it seems kinder to sneak them out of the habit by progressively diluting their milk feedings with water.

Sometimes a calf that was thought to be weaned begins sneaking a lunch out there in the pasture. You can fix that quickly with what's called a *calf weaner* or *sucker,* a bit of spiked hardware that fits in the calf's nose and so jabs any cow he or she tries to nurse. It doesn't take many retaliatory kicks before the calf learns that milk is no longer on the menu.

Calf Housing

The best home for a calf is about a five-by-five box stall with solid or slatted sides, plenty of dry bedding and no drafts. Even an infant calf can stand cold as long as it's dry and sheltered from gusts of wind.

If there is no box, the calf may be tied by leather collar and short rope to some sheltered nook in the barn. Again, it should have lots of bedding, preferably straw or hay into which it can nestle.

A strong and healthy calf can go outside for a few hours of exercise any time as long as the weather is suitable. It will learn quickly what a tether is all about. Because it is more susceptible to worms and other parasites that attack cows, a young calf often is kept out of pastures where adults have been running, until it is at least six months old.

The calf lot or wherever the calf is tethered must have some shade on a hot day. And the calf shouldn't be left for hours without an offering of water.

Horns

A calf that isn't polled is born with little horn buttons just beneath the skin of its forehead. If the calf is going to be dehorned the best time to do it is before these buttons sprout. Most vets recommend that it be done with caustic soda or a prepared dehorning compound when the calf is one to two weeks old. Since the directions are a little different for each of these compounds, it's best to buy what is available at the nearest farm supply store and follow the recipe.

Those who like horns and who want to get fussy about their shape can get help from the Jersey or Ayrshire associations on how to train them. This process, done with strings or weights or both, is begun later when the horns have grown into three- or four-inch spikes.

Castrating Bull Calves

Bull calves that aren't wanted for breeding should be castrated when they are three to six months old. Some say it can be done when they are but three days old, but this seems too early. Even if the testicles have dropped into the scrotum it would be better to give the calf at least a couple of weeks' start in the world before putting him through the operation.

The way we have done it does not seem to bother them for more than a few minutes. We use what is called an *emasculatome*—an instrument that squeezes down on either side of the scrotum crushing the cords and arteries within while not cutting the outside skin. This is a bloodless operation, as is the use of hard rubber bands with the instrument called an *elastrator*. You needn't worry about flies or infections, as you must whenever a knife is used to cut into the scrotum and remove the testicles.

The best instructions on castrating will accompany whatever instrument is bought or borrowed for the job. Also, castration is pretty well described in several amateur veterinary manuals, one of which is listed in *Other Books and Places* in the appendix.

Young cows are naturally playful. They love to butt heads or push against a hand held to their forehead. It's not a good idea to play this way with a bull calf that is to be raised for breeding. It gets the wrong pattern of ideas going for everyone concerned.

The way to treat them all is with gentleness and a firm hand. Put them in a halter early on and teach them to be led while they are still small enough to control.

Heifer Calves

Heifer calves being raised for production will benefit from a pound or two daily of a 16-percent CP concentrate any time up to four months before calving, except perhaps at the times when they are on unlimited good pastures or excellent dry roughages. Use your own judgment about raising or lowering the protein value of the concentrates, depending on the roughage quality.

A mineral mix as suggested in Chapter 11 would be especially helpful for these rapidly growing animals, unless perhaps their hay or other forages are around 50 percent legumes or their diet includes some other high-calcium foods—grown on well-fertilized soils to guarantee the phosphorus side of the equation as well.

As the bred heifer comes within four months of calving (she now may be called a "springer"), she will benefit from a gradual increase in concentrates to a level of five to seven pounds a day. Watch her udder for swelling (edema), and if she becomes obviously uncomfortable or begins to leak milk days before she is due to freshen, cut back on the rich foods.

CHAPTER 17

Health

The best insurance against losing a cow or having to spend huge amounts of time or money treating an illness is to know how she behaves when everything is going well. That way you can recognize problems before they become difficult or impossible to treat.

The owners of family cows have an edge over large commercial farmers because they can get to know their animals as individuals. The edge is sharpened through the close personal attention that automatically comes with hand-feeding, cleaning and milking.

Know Your Cow

In getting to know your cow, use all of your senses. Learn how she moves, how she looks, feels and sounds. Get to know the normal smells of the barn where she's kept. Smell the cow's breath. Listen to her breathing. Watch how she eats. Watch how she normally reacts to a slap on the rump or to the normal confusion and noise in the barn. Buy a veterinary thermometer early in the game and get to know your own cow's range of temperatures through an ordinary day.

Use a girthing tape or a measuring tape and the table in the appendix to estimate the cow's weight. Overall the cow should gain weight until she reaches full size at about the age of two to two-and-a-half years. From then on, her weight should remain about constant except for losses of up to 100 pounds or so early in each lactation.

Although I am leery of drugging every bug, and think sometimes that veterinary medicine is too oriented in that direction, I do look to the vet to back up a preventive health care program based on the idea that cows want to be healthy and will be healthy if they are eased along in an environment of clean, dry pastures and barns and ample good food.

Veterinary Help

In spite of preventive efforts, things do go wrong. The question then is when to call in the vet. Most vets willingly respond to emergency calls at any time of the day or night, with two exceptions. The first is when in fact there is no emergency. The worst is when there is an emergency that wouldn't have occurred had the owners called for help a little sooner.

WHEN TO GET IMMEDIATE HELP

Here is a list of some signs that call for immediate attention:

1. Heavy and uncontrolled bleeding.
2. Cow down and on her side. Unable to get up or roll over. (Could be milk fever if this paralysis comes within a few days of calving. If it *is* milk fever the cow's temperature actually will be a bit lower than normal.)
3. Cow has gone into labor but has given up. The calf may or may not be partly visible. (See Calving Problems in Chapter 16.)
4. Prolapsed or everted uterus following calving. A partially retained afterbirth, which is serious but not immediately critical, may look something like a prolapsed uterus. But the remains of afterbirth not fully expelled will be more like dangling ropes, while the prolapsed uterus is definitely bulbous and up to the size of a medicine ball.
5. Cow up or down, but immobile unless prodded, and not chewing her cud. If she appears blown up she may be suffering from bloat.

6. Temperature of 104° F. (40° C.), especially when coupled with other abnormal conditions.

7. A sudden drop to a subnormal temperature in a cow with an obvious infection or one that has been under treatment for a serious illness, but that shows no other signs of improvement. She may in fact be going under.

8. A calf with white or bloody, watery and reeking diarrhea, or a calf with any type of diarrhea that has also stopped eating.

WHEN TO GET HELP SOON

Here is another list of some of the signs that call for attention within the next 12 to 24 hours:

1. Cow up or down, immoble, but chewing her cud. A cow that is on her side must be rolled from hip to hip as often as possible to keep gases from becoming trapped in her rumen.

2. Cow has refused all water and food for the past 12 hours.

3. Cow has refused most food, including concentrates for 24 hours or more.

4. Temperature up two or more degrees above normal for the time of day and activity, again especially when coupled with other abnormal conditions or behavior such as a limp, swelling, odd smell or general lack of energy or appetite.

5. Cow or calf that has had a prolonged bout of diarrhea that can't be blamed on a change of feed—such as a cow on flush pasture—and especially if the animal's skin is losing its elasticity and/or the eyes appear sunken. These last are signs of serious dehydration, and recovery may only be possible with injections of fluids.

6. Bloody urine or feces.

7. Persistent coughing, hiccupping, shivering or labored breathing—in fact, persistence of *any* gross abnormal behavior that interferes with the cow's normal routine of drinking, feeding and resting.

8. Hard swelling developing in the udder, more easily noticed after the disappearance of normal swelling and caking that goes with calving.

9. Any obvious change in the appearance or consistency of milk.
10. Any sudden and obvious drop in milk production that persists beyond one milking. (Check water supply. Or have there been other major changes in the cow's room or board? Numbers 8 and 9 and 10 could be signs of udder infection, *mastitis*.)

First Aid

Every cow owner should be able to give some first aid and nursing service. A wonderful way to pick up on many of the skills is through making the rounds of neighboring farms with the local large animal vet. It won't hurt to ask if it's possible. Most vets seem to enjoy company, and the ones I have known have been born teachers. Maybe it's part of being a practicing rural vet where farmers are often the vet's only assistants.

In any case, here are a few suggestions on taking temperatures, giving pills (boluses), shots and drenches, and also on stitching and bandaging superficial wounds.

TAKING TEMPERATURE

Although a dry muzzle often indicates a fever, a far better reading of a cow's health can be had by taking her temperature with a rectal thermometer. Use a veterinary type with a string tied to the back end.

Most dairy cows' body temperatures range from 101.3° to 103.1° F. (about 39° C.). After feeding, running or standing in the hot sun their temperatures will tend to be higher. But no two cows are exactly alike, and so an individual record of temperatures through a normal day is an advantage when it comes to determining one cow's health, the best time for breeding and, possibly, when she is about to calve.

To take a cow's temperature lift her tail and slowly insert the oiled or Vaselined thermometer to its hilt within her rectum. Leave it there for three minutes at least. Veterinarians can afford to leave a thermometer untended in the cow's rectum. I'd rather waste three minutes holding onto the string or the tip of the thermometer than to have to chase after one that was sucked within or got pushed into the gutter.

INJECTIONS

Most shots or injections are divided into three groups depending on where the drug or chemical is injected. *Intravenous* (IV) injections put medicines directly into the bloodstream. *Subcutaneous* ("sub cue") puts them under the skin, and *intramusculars* (IM) go into muscle tissues. Before giving any injection, clean the target area with alcohol or with water and soap.

IM is the simplest because all it takes is a slapping insertion of the needle into any well-fleshed spot. The cow's thigh is commonly used. Give the cow warning that you are there and give her a slap or two with the heel of your hand before driving the needle home.

Most IM drugs must not go directly into a blood vessel because they could cause clotting, destruction of blood cells or even death. To make sure no vessel has been hit, draw back on the plunger of the syringe before injecting the drug. If the tip of the needle is in a vessel the syringe will draw blood. If this happens move the needle to a new location.

With "sub cue" injections lift a piece of the cow's skin and carefully insert the needle just under the fold. Imagine that you are injecting a rubber ball. You will feel when the needle has pushed through the cow's tough outer hide.

IV shots are the trickiest because it takes finding the location of a large blood vessel near the surface. The jugular vein on the left side of the cow's neck is the usual target for these shots.

After tying the cow's head securely with a halter or nose-lead (using the quick-release bow knot shown in Chapter 5), tie a rope tourniquet around her neck by her shoulder. This should lift the vein into view above the rope and in what is called the *jugular furrow*. With IV injections you also want to pull back on the plunger or at least allow it the freedom to move back showing blood in the syringe; proof that the needle is where you want it. Again, as in IM shots, give the cow a slap or two before inserting the needle. Release the tourniquet before injecting the fluid or drug.

PILLS

An oral pill or *bolus* is easiest given with the help of a balling gun that places the pill far back in the cow's throat. A home-made balling

Balling gun showing pill or bolus within.

gun can be made out of a piece of pipe or plastic hose about 18 inches long, with an inside plunger an inch or two longer than the outer tube. Often a cow will accept and swallow a bolus better if it's sparingly wrapped in a bit of tissue or toilet paper.

Hold the cow and run the balling gun back along the gums until she opens her jaws. Then push gently but resolutely to the back of her throat before pushing the pill from the gun.

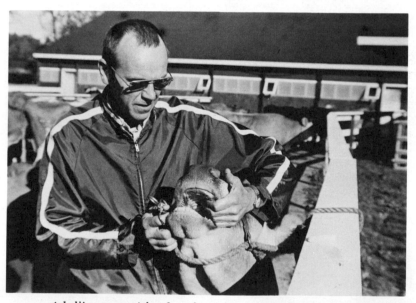

A balling gun, either bought or homemade, is a must when it comes to giving a cow a pill (bolus). Make sure that the cow is tied securely by a rope or stanchion before you begin, working with the tip of the balling gun back along her gums. When the cow opens her jaws slowly work the gun to the back of her tongue. When the pill is dropped far enough back in the cow's mouth it is almost impossible for her to avoid swallowing it down.

GIVING DRENCHES

Drenches are fluids given orally, usually for treatment of some dis-
order in the digestive system, such as bloat or intestinal parasites. They
can be given with any long, narrow-necked bottle. Or a squat bottle
can be used if it is fitted with six or eight inches of rubber or soft plastic
hose. Rubber teat-cup liners are often used this way.

Hold the cow's head as in giving a bolus but be careful to hold
her muzzle only slightly above a line parallel with the ground—only
enough above parallel to let the drenching fluid trickle back to the cow's
throat. Insert the bottle neck or tube along the cow's gum and pour a
little fluid. The cow will fight and mouth and chew, and hopefully
swallow. If you're using a glass bottle without a soft rubber or plastic
hose make sure to keep the glass from between the clamping jaws. Pour
the fluid only a shot at a time, giving the cow time between sloshes to
swallow. It may help to have someone standing by gently stroking the
cow's gullet between tips of the bottle. This stimulates the swallowing
reflex.

Drenching takes the most care and skill because one wrong move
can be fatal. I've drenched cattle and sheep many times without a
hitch, but I am always taking a chance that some day I may goof. All it
takes is for the cow to breathe at the wrong instant, and she could go
down with her lungs filled with fluids. If you've any fears, call in a vet.

TREATING CUTS

It's a good idea to have a hefty roll of four-inch or five-inch gauze
bandage around in case a cow or calf gets cut or needs to have an infected hoof wrapped. If there is no gauze, use any clean cotton cloth.
In bandaging a hoof, foot or leg, wrap the bandage around and around,
between the hooves if necessary, then cut off the end of the bandage
and tear back, leaving two ends that can be tied around the injured
limb to secure the wrap. The bandage has to be snug enough to hold in
place but not so tight that it cuts off circulation. Leave just enough room
to allow a finger or two to be slipped between the bandage and cow.

To keep a bandaged or medicated hoof clean, tie on a boot made
from an old innertube or two or more layers of woven plastic feed bag.

Flesh wounds that tend to lie open often can be stitched together as easily as a hole in a pair of pants—sometimes without novocaine, since cows and calves generally are more tolerant of the pain of a sewing needle than we are. If the wound is in a sensitive area and there is no professional help available, the animal may have to be taken off her feet and tied. (See Chapter 5.) Locking forceps, available through many livestock supply houses, help to hold the needle. (These forceps also are the only reasonable tool to have around for pulling porcupine quills out of dog or cow muzzles.) Clean all wounds thoroughly and flood with peroxide, iodine, Merthiolate, alcohol or liquor before closing. Alcohol-based disinfectants are harsh on open wounds, but if that's all you have, it's probably better to use one than to leave the wound untreated.

Thin-skinned calves are easiest to stitch. Any large sewing needle will do and the thread may be whatever toughest thread—preferably nylon or some other synthetic—is available. Clean the thread by boiling it in water or by soaking it for a few minutes in alcohol, hard liquor (high-proof gin, vodka or light rum are best) or iodine.

Cow hide is tough and it is best to have large, curved needles on hand for the job. Curved upholstering needles and needle-nosed pliers in place of locking forceps would be OK stand-ins in an emergency.

Close a wound using separately tied stitches an inch or less apart. More stitches are needed for skin that has to stretch when the cow moves. Run the needle and thread through the skin on each side of the wound at each stitch site, then tie the ends of the thread together in a tight knot. A broken, sterilized rubber band laid in a deep wound and with one end protruding may help with drainage in case a minor infection develops.

The stitches and rubber band can be removed after a week or so as the wound appears healed. Snip each knot with the tip of a pair of scissors and gently pull the thread free.

I would like to go into health problems more thoroughly with symptoms, causes and cures for the most common, but it's too big a subject, needing books of its own. The decision wasn't easy. The bed, floor and chairs are covered now with charts and lists and letters from local and federal animal health people: common diseases; the most costly, big problem in the Southwest; and on. Where to draw a line, because there isn't room for it all?

One line makes more sense for this book because it leaves me with two diseases that are way out ahead of all other health problems by being the big destroyers of milk production and calves. One is *mastitis,* or *garget,* and the other is *calf scours.*

Both mastitis and scours are collective words describing symptoms brought on by many different bugs or poor physical conditions. Maybe that's one reason why they seem more important than their tiny names would say. *Mastitis* means inflammation of the udder or mammary system, and it can be caused by a number of microscopic organisms given a helping hand by filth, cuts, cold, bruises or anything that puts the milking cow under stress. *Scours* is another word for diarrhea, a signal that something has gone wrong in the large intestines. Some types of scours, the ones brought on by bacteria or viruses, are contagious.

MASTITIS

Mastitis is always contagious. The two most common culprits are *staphylococcus* and *streptococcus* bacteria that enter the udder through an injured teat or up the teat canal past weakened muscles and/or secretions that are supposed to keep the end of the teat sealed.

A cow may have a low-grade or chronic mastitis infection that rambles on for months without being noticed unless a person keeps a close watch for slight changes in the milk or has a sample of milk sent to a pathology laboratory for testing. But anytime, and without there necessarily having been a chronic infection brewing along, an acute or severe attack of mastitis can flare up. The most severe attacks can affect the whole cow, bringing on a fever and lowering her resistance to other diseases. Even if the cow recovers her general health after a very severe attack of mastitis, there is a good chance that one or more quarters of her udder will have been permanently ruined and blocked with scar tissue.

Controlling mastitis takes cleanliness and careful housing to prevent teat and udder injuries. It will be more difficult with older cows whose pendulous udders may leave their teats in a hoof's way; whose teat muscles have begun to lose their tone, and whose general ability to resist infection is down from the level in young, robust animals.

Control takes a careful eye for the least sign that there is something abnormal about the milk. Usually mastitis attacks one quarter at a time. If it is caught early enough it is usually easy to knock it down with antibiotics, or maybe all it will take is some careful milking for a day or two.

The first signs of abnormal milk will best show with the use of what is called a *strip cup* with a finely-meshed screen that takes the first squirts from each teat. Minute flecks of white will gather on the screen warning that something is wrong in that quarter. In place of a strip cup you could use a piece of smooth, black plastic. Squirt the first milk across the plastic and onto the floor. Look for the white flecks against the blue-white milk. Or, direct the first squirts from each teat against the side of the pail, where the flecks may show.

Strip cup

With a severe attack of mastitis there will be stringy clots in the milk, especially at the beginning of milking. As the attack worsens the clots may become bloody.

Whether antibiotics are used or not, treatment includes milking the infected quarter(s) last to keep down the likelihood of spreading the bacteria to the uninfected quarters. It takes cutting back on concentrates so that the cow and her udder are under less pressure to produce. And it takes milking the infected quarter(s) very thoroughly with gentle massaging to make sure that all of the milk is drawn and that blood is circulating well through all of the lactating tissues. Following milking, all of the teats should be dipped in a specially-prepared disinfectant solution. Regular iodine containing phosphoric acid is too strong.

Antibiotics are usually squeezed into infected teats and quarters through a plastic tube inserted up the teat canal. When they are used antibiotics pass into the cow's circulatory system and this way get into the milk in uninfected quarters as well. All of the cow's milk will be temporarily contaminated—spoiled for cheese-making and for people who are allergic to these drugs.

SCOURS

Calf scours is the big killer of young stock. It usually hits within two to three weeks of birth. The worst type is often described as the *white scours*. It's white to yellowish, and it smells worse than cat scats. The only difficulty for beginners is that the first manure anyone sees from a calf is yellow, pasty and smells foul. Yet that's normal.

No matter what the color or smell of it, a very loose and watery manure is the thing to watch for and react to immediately. The small calf has no body reserves to hold it over while foods roar unused through its system.

Fortunately, most bouts of calf scours are nothing more than signs of a digestive upset brought on by overfeeding or the feeding of dirty or cold milk. If the calf is being bucket-fed, cut her allowance of milk in half or more while making up the fluid difference with warm water to prevent dehydration. If the calf is too young to be eating or drinking well, try carefully to get warm water down its throat with a basting syringe. If the diarrhea is no better a day later call the vet. If the calf does come around, gradually increase the milk over the next couple of days until she is back to recommended levels.

The color and extreme stench of "white" scours should be the clues that here is something that wants a vet's opinion or treatment within the next half day. A fever may not be noticed with scours. The fact is that normal or sub-normal temperatures are more often noted and may be signs that an infection has gained the upper hand. Antibiotics will help fight the initial infection and help prevent an invasion of bacteria that would be likely to step in while the calf's resistance is down.

CHAPTER 18

Keeping the Land

For most people, part of keeping a cow is going to be keeping the land that feeds her. If the land isn't kept it will soon revert to the weeds, shrubs and trees that belong with the area's climate and soils.

Some of the wild grasses and weeds aren't bad foods. It's only that in many cases they don't produce as much food as could be had with cultivated varieties because they are low-growing, early maturing or have some other peculiarity that makes them hard to use.

None of our major livestock crops can make it alone in North America. Most came from Asia, Africa or Europe. Even those whose ancestors did originate here—corn and potatoes—have been under cultivation for such a long time they can't survive without human help.

So how is it done, keeping five or ten acres the way nature never intended? Working part time and with only a cow and a few other animals to support the venture makes it seem almost too big a job to take on.

Rotation and Improvement

The important idea to keep in mind is that it doesn't have to be done all at once. The four-acre family mentioned in Chapter 1 keeps its cow and land by improving one small piece every year in a rotation that includes their vegetable garden.

These people have no permanent pasture, but their one big hay field is fenced. In the early part of the pasture season the cow is tethered along the driveway and road and around the barn and outbuildings—wherever there is a day's worth of grass.

Meanwhile, inside the hay field all of the previous winter's manure and some lime have been spread on the new year's garden ground. Last year's garden is plowed and seeded to a combination: a small grain like oats, plus timothy and red clover. (It could be alfalfa or whatever legume suits the climate and soils.)

After haying in late spring the cow is moved onto the field. A temporary electric fence keeps her out of the garden and newly seeded ground. In late summer the newly-seeded ground provides a soiling crop—mostly green oats and clover. The next summer it gives a top-quality cutting or two of timothy and clover. From then on timothy is the dominant crop until it comes time for that piece of ground to become garden again.

Liming

Whether it is done by stages or all at once, most cleared land in North America that hasn't had any attention for several years could probably stand a heavy application of lime. This could be as much as three or four tons to the acre if you are living in an area that gets more than 30 inches of precipitation a year. The lime will increase overall growth and encourage better forages like the legumes (vetches, clovers, etc.) that want a less acid soil.

There are two main reasons why acids build up in the surface of warm, moist soils. First is that more acids than bases result from chemical reactions that go on as organic matter decomposes. Rain and snow melt percolating down through the soil carry more bases than acids beyond the reach of roots. This is because clay and organic matter in topsoil have a greater ability to hang onto the acids. Eventually the topsoil becomes so saturated with acids that crop roots are starved for basic minerals they need—like calcium and potassium or magnesium. Also an acidic condition may bring about a situation where other minerals such as aluminum are brought into the soil water in quantities and/or chemical forms that are toxic to numerous species of plants.

Trees and other deep-rooted plants can bring their basic needs up from the subsoil. Many wild grasses and weeds are adapted to high-acid conditions. For these and other reasons many of our agricultural forage crops, particularly those with shallow roots and low acid-conditions tolerance, can lose out to perennial weeds, shrubs and trees unless they are given periodic boosts of basic minerals from above. Theoretically several basic minerals could serve the purpose, but lime (calcium carbonate) is best, plentiful and relatively cheap.

Although an estimate of soil acidity and lime needs can be made with a pH testing kit, it is better to get a complete soil analysis through an agricultural department laboratory, because soils of different textures, compositions, and under different climates need different amounts of lime to counteract the effects of accumulated acids.

pH is a measure of the intensity of acidity in soil water. The pH scale runs from 0 to 14, with 7 being neutral. Any reading below 7 is considered acid. Above 7 is basic, or alkaline. Each number on the pH scale represents a ten-fold increase or decrease in acidity or basicity (alkalinity) depending on which way you run. So a soil with a pH of 6 is ten times as acid as one with a pH of 7. And a pH of 5 is 100 times more acid than 7.

Untended agricultural soils in humid regions of North America often will show a pH of about 5. This is OK for potatoes. But most vegetables and domestic forages will want a pH of about 6.5.

SOIL SAMPLES

Agricultural department soils tests are better made six months to a year before planned improvements. This gives the busy labs a chance to get the results back in time for you to buy what is needed and prepare the attack.

A pint-sized soil sample should come from a mix of several surface-to-shovel-depth cores of dirt taken from different corners and elevations in a garden or field. Otherwise it won't give an average picture of what is going on.

Plants aren't as sensitive to overliming as they can be to an overdose of chemical fertilizers. It can be spread by shovel from the back of a truck or wagon. Usually lime is spread a few months before planting or the growing season to give it a chance to work. Do it anytime the ground isn't frozen, and, best but not essential, follow up with plowing or disking so that the lime won't wash off the land.

Fertilizers

Chemical fertilizers providing nitrogen, phosphorus and potassium must be spread with more care. Otherwise the soil solution may be overloaded in spots with nutrients that can be toxic in high concentrations, or that form such salty environments that nearby roots are dried like fresh sides of bacon packed in salt. Sometimes heavy applications of commercial fertilizers are plowed in before a piece of ground is reseeded. Or else lighter applications are added to established fields or pastures as a "top" or "side dressing."

It is impossible to suggest specific rates of chemical fertilizer applications since there are wide variations with different soils, crops and fertilizer grades. Usually it will be a question of hundreds of pounds to the acre, not tons. It's better to fertilize after liming, so that the new minerals added don't get tied up with acids and made unavailable to the plants. Also there will be a better return from land that has been or is going to be seeded with improved crops.

Mixed, complete commercial fertilizers come coded, like 5-10-5, or 15-10-10. The first number is almost always for the percent of nitrogen in the bag, the second for the percent of available phosphorus (as phosphate P_2O_5), and the third for the percentage of potassium, or potash (K_2O).

Because legumes fix nitrogen for themselves and neighboring plants, fields or pastures having 50 percent or more of these crops will make best use of a fertilizer low in this nutrient. Grasses, grains, roots and most of the vegetable crops will want nitrogen unless they are planted on ground recently in one of the legumes.

And so it is unlikely that the needs or conditions of a soil for these three basic mineral elements will be exactly the same. And it would be best, as in liming, to work from a crop plan and on the basis of a professional soil test.

If there is no time for a soil test and yet you want to get some minerals on the land it won't hurt to apply up to 500 pounds to the acre of a 10-10-10 or 10-15-10 fertilizer, even on land that is getting plenty of compost or cow manure.

The liming at three tons per acre should be sufficient to hold the proper pH for about five years. There may well be a need for more nitrogen, phosphorus or potassium within a year. But don't go on year after year applying fertilizer blindly.

Fertilizers can be spread evenly over the land and always should be for pastures or hay fields. For row crops it can be trickled and buried in shallow trenches a couple of inches to the side of the seeds.

Keep livestock off freshly fertilized fields or pastures for at least a week or until rain has washed the forages. If it rains, wait for the puddles to dry before putting the stock out where they might drink the chemically charged water.

Manure

Although the nitrogen, phosphorus and potassium are most talked about, well-handled manure brings quantities of important organic materials, microorganisms and trace minerals back to the soil.

A medium-sized dairy cow produces about 13 tons of manure a year (feces plus urine). About half will get an automatic spreading around the pasture. Roughly this is equal to putting six and a half 100-pound bags of a 10-5-10 fertilizer on the pasture and can be considered when planning how much of other fertilizers to spread.

The other six and a half tons of manure (figure 50-70 pounds a day through the barn season) will have to be spread from a wheelbarrow, wagon or small automatic spreader resurrected from an old barnyard. Its fertilizing value will depend on how well it is kept and treated.

Ideally all feces and urine would be collected, spread and plowed or disked into the ground daily, with the anticipated year's supply being distributed evenly over all of the crop lands. The next best treatment would be to have a water-tight cistern into which everything ran or was shoveled until the time came to move the gloop onto the fields. Next best would come periodic spreading from a well-packed pile kept under cover.

Leaving manure piled against the barn, washed by the rain and eventually spread months later, is the worst way to treat the barn and the manure. Wood rots fast against a bank of manure. Most of the manure's nitrogen will be converted to gases that are blown away, or into soluble nitrates that get leached into the soil. Many of the other

Birdsfoot trefoil

minerals will be washed away too, leaving a lush patch of ground by the barn and starved fields.

A quantity of moist, acidic manure, denied air, ferments and is preserved like silage. And so a practical way to handle manure on a small place is to first of all use plenty of bedding—sawdust, shavings, peat moss or straw—to sop up the urine. Then spread the manure as frequently as possible, from a well-packed and moist pile sheltered by a roof in a humid climate or built straight-sided and dished like a fallen cake. This gives the rain less of a chance to wash away the soluble nutrients.

The best time to spread manure is just before it's going to rain so that the nutrients will wash where they're wanted.

A periodic sprinkling of dirt or peat moss will help hold the nitrogen in an outdoor pile. Another method is to throw a pound or two daily of a 20 percent superphosphate fertilized in the gutter (10-20 pounds of P_2O_5 in the form of superphosphate per ton of manure). This reacts chemically to hold the nitrogen, plus adding to the low phosphorus level of manure.

Although usually expensive, peat moss makes a fine bedding material because it absorbs more urine per pound than other common bedding materials. It is high in nitrogen and is strongly acid. This last, as

SOLID AND LIQUID EXCREMENT OF
MATURE ANIMALS. AVERAGE DAILY
AMOUNTS AND COMPOSITION

COMPOSITION OF FRESH EXCREMENT

Kind of Animal	DAILY PRODUCTION PER ANIMAL		DRY MATTER		NITROGEN		PHOSPHORUS*		POTASSIUM**		LIME (CaO)	
	Solid (lbs.)	Liquid (lbs.)	Solid (%)	Liquid (%)	Solid (%)	Liquid (%)	Solid (%)	Liquid (%)	Solid (%)	Liquid (%)	Solid (%)	Liquid (%)
Horses	35.5	8.0	24.3	9.9	0.50	1.20	0.30	Trace	0.24	1.50	0.15	0.45
Cattle	52.0	20.0	16.2	6.2	0.32	0.95	0.21	0.03	0.16	0.95	0.34	0.01
Sheep	2.5	1.5	34.5	12.8	0.65	1.68	0.46	0.03	0.23	2.10	0.46	0.16
Hogs	6.0	3.5	18.0	7.3	0.60	0.30	0.46	0.12	0.44	1.00	0.09	0.00
Hens	0.1		35.0		1.00		0.80		0.40			

* Phosphoric acid (P_2O_5)
** Potash (K_2O)

(From USDA Yearbook of Agriculture, 1938.)

in adding superphosphate, helps create the right conditions for acid fermentation in the manure pile.

With any luck a ton of cow manure handled this way ought to be about equal to 100 pounds of a 3-5-10 fertilizer (or a 3-10-10 fertilizer if the superphosphate were added)—in either case, a perfect fertilizer for legume crops. If, when the manure pile is opened, it's white and dusty inside, it means the air-loving molds and other microorganisms of rot have thrived. Much or all of the nitrogen will have been lost. Keep the next pile trampled down and maybe add a little water. It's hardest to avoid this "fire fanging" when using lots of straw or old hay for bedding because the manure doesn't want to pack down.

Seeding

In land improvement the first thought is lime, then fertilizer, then seeding. The choice of grass or legumes depends again on climate and soils. The accompanying map suggests possibilities around the United States and Canada. Of all the imports timothy is the hardiest over the widest area. "Volunteer" stands are common, showing that it seeds itself and competes fairly well with native grasses and weeds.

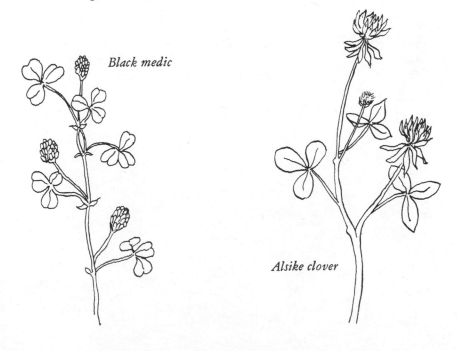

Black medic

Alsike clover

MOST ABUNDANT FORAGE GRASSES AND LEGUMES BY REGION

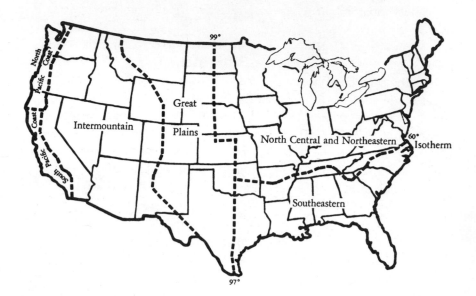

NORTH CENTRAL & NORTHEAST

Grasses	Legumes
Kentucky bluegrass	white clover
timothy	Korean lespedeza
redtop	sweetclovers
orchard	alfalfa
Canada bluegrass	common lespedeza
tall oatgrass	red clover
meadow fescue	alsike clover
smooth brome	hop clover
sudan	black medic
ryegrass	crimson clover
bents	birdsfoot trefoil
beardgrasses & bluestems	

GREAT PLAINS

Grasses	Legumes
grama	sweetclover
wheatgrass	alfalfa
buffalo	
brome	
galleta, tobosa, curly mesquite	
bluegrass, bluestem	
sudan	
bluegrass	
timothy	
redtop	
fescue	
Rhodes	

SOUTHERN

Grasses		Legumes	
Bermuda	redtop	common lespedeza	black medic
carpet	Bahia	hop clover	spotted bur-clover
dallis	fescue	white clover	
vasey	Rhodes	Persian clover	

220

INTERMOUNTAIN		NORTH PACIFIC COAST	
Grasses	*Legumes*	*Grasses*	*Legumes*
wheatgrass	alfalfa	ryegrass	red clover
grama	white clover	Kentucky bluegrass	white clover
brome	sweetclover	bent	hop clover
galleta, tobosa,	alsike clover	reed canary	alsike clover
curly mesquite	red clover	orchard	
bluegrass	black medic	meadow fescue	
timothy	California bur-clover	redtop	
redtop	strawberry clover	timothy	
Bermuda		tall oatgrass	
beardgrass or bluestem		meadow foxtail	

SOUTH PACIFIC COAST	
Grasses	*Legumes*
fescue	California bur-clover
brome	alfalfa
wild oats	white clover
Bermuda	black medic
sudan	

(From USDA *Yearbook of Agriculture*, 1939.)

Alfalfa is one of the best of the legumes, although in the North freezing weather usually kills it back within three to four years. Red clover is a biennial. Unless it can reseed itself it is only good for two years.

If an old field or pasture hasn't gone too far to weeds it may be worthwhile liming, fertilizing and harrowing a piece in the fall followed by a broadcasting of seeds in early spring. Because the seedbed has only been chewed over, increase the usual rate of seeds per acre by 25 percent or so. For small acreages those seed broadcasters that hang by a shoulder harness will do an excellent job.

Cultivating

Many old fields are so far gone to weeds and shrubs that the only recovery is through plowing. This is rough, hard work, especially in well-established sod laced with woody roots. In most cases it will be cheaper to hire the plowing done than to buy even the cheapest single point (bottom) tractor plow. There is one exception: the old horse-drawn plow, if one can be found. Pulled by a horse, tractor or 4-wheel drive car, it will turn the sod handily—and the operator too if you hit a rock too hard, which is the only drawback. Some of the oldest plows still around have wooden "beams," the bar running from the top of the plow blade forward to the hitch. It's better to find one with a steel

From the looks of things, the plow above has fetched up on her last rock. A shame to see it. There can be miles of useful service in old tools, as the people below discovered when they hitched a plow of the same vintage behind their jeep.

beam, and the best of these newer models have more versatile two-way blades.

Long grass makes plowing more difficult. Unless you're using tough, modern equipment, it will probably help to mow off a field before attempting to turn the sod. It will be easier plowing in spring or at least at a time when the ground is moist—but not soggy. Clay soils especially can be injured through being worked when they are too wet. They tend to pack and then dry into rock-like clods. To test a soil for proper moisture level take a small fistful and squeeze it into a ball. If the ball does not crumble apart with ease, the ground is too wet to work.

After plowing, the sod has to be broken up. A disk harrow is the best tool, but if you haven't got one go directly to a spike or a spring toothed harrow. If you haven't either of these, tie a ten-foot, fat and branchy log (with the branches cut off about a foot from the trunk) on an angle behind whatever animal or vehicle can tow it over the plowed ground.

Harrow plowed ground until the sods are broken and leveled, but don't get too particular. Driving over a field packs the soil down, leaving less space for air and water around the soil particles. And a surface that is too smooth won't provide the best bed of nooks and crannies for the germination of the broadcast seeds.

Most of the forage crops can be mowed for hay or soilage six to eight weeks after planting. Given the right climate and soil fertility

Disc harrow

some may give two crops in the first year in the North, four or more in the South, with irrigation.

Ground that is going to be planted to roots, corn or some other row crop will have to be better prepared so that the land can be cultivated after the crop is up. Any amount of large sod gets wrenched by the tines of a cultivator, and the young plants are uprooted and buried. Hoeing, too, if that's the way it's to be done, is twice (or worse) the job when there are large hunks of sod between the rows.

CHAPTER 19

Making and Storing Hay and Roots

Hay made on a small place will be sun-dried grass or legumes—maybe a mixture. The process of making it is simple, if you get the sun. Some years the sun and the haying come easily. Other years there are lots of discouraging words as the weather forecasters goof and new-mown hay turns to compost.

If there isn't the time or you can't stand the possible grief, either buy hay or maybe help a neighboring farmer make hay in exchange for enough bales to see you through the winter.

Making Hay

The time for mowing is a time of compromise between the best quality and the most quantity of hay. An early cut means better quality hay but less of it—although in northern climates the earlier the cutting the more chance there may be to get in a second crop. A late cutting beyond the time of best compromise fills the barn with more tons of hay but it will be a tougher, lower protein crop.

The graph included here shows when timothy, bromegrass, orchard grass and alfalfa are in their prime for making hay. This is fine to keep in mind—but the reality of haymaking is not always so clearly cut.

WHEN TO CUT

For one thing, few fields the family cow owner comes by will be uniform. They will have some or mostly mixed wild grasses. Deciding when to mow these fields won't be the simple matter of looking out the window one morning discovering that, "Hey! The timothy's in the 'boot.' Time to do it."

EFFECT OF MATURITY ON DIGESTIBILITY OF TIMOTHY, BROMEGRASS, ORCHARDGRASS, AND ALFALFA

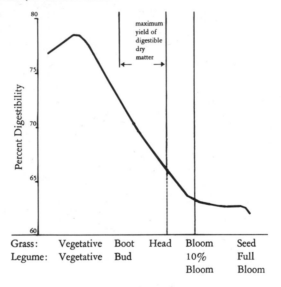

STAGES OF DEVELOPMENT WHEN CUT

NUTRITIVE VALUE CHANGES AS CROP MATURES

Stage of Growth	TDN %	Crude Protein % Grass	Legume	Intake Value Lbs./100 lbs. Body Wt.
Vegetative	63	15.2	21.0	3.0
Bud	57	11.3	16.4	2.5
Bloom	50	7.5	11.5	2.0
Mature	44	4.1	7.3	1.5
Regrowth	52	11.3	16.1	2.6

(From "Field Crop Bulletin," from the Atlantic Provinces.)

The only answer is to cruise the fields from time to time, judging the age and best time to cut on the basis of what is happening to one or two dominant species. Hope to get mowing within a week of the first signs that they are heading into bloom.

There is no sense working up a froth over any precise time to get started. The weather will always have the last word. Late hay is better than no hay at all. More of it will have to be fed, more frequently, so that the cow can pick through for the best and not be forced by hours of fasting into cramming her paunch full of this less-easily-digested roughage.

Whenever it happens, the best hay is made by rapid drying and a minimum of handling. Fast drying stops chemical reactions that can convert good dry matter into gases that are lost. Overhandling hay that is almost ready for the barn causes losses through "shattering" of the most valuable leaves and leaf-tips. Legumes are harder to cure because before the lush stems are fully cured the tiny leaves are likely to be fully dry and ready to flake off.

Overly dry and weathered hay of any kind shatters badly. And yet if loose hay is stored before it is down to 27 percent or less moisture, or baled or chopped hay is stored before it is even drier—22 percent moisture or less—then there is a good chance for heating and rotting. There is the possibility of fire through spontaneous combustion if, as occasionally happens, fermentation and molding give way to rapid oxidation.

Tools for Haying

Modern hay-making equipment costs thousands of dollars. But it's not needed for making the amounts of hay needed to winter a cow and calf. A combination of simple tools could easily cost less than buying one year's supply of hay. All that's needed is something to cut with, something for raking, a fork to gather and stack and a truck, wagon or cart to haul the hay to the barn.

The simplest combination begins with a scythe, then a wooden rake, a pitchfork and a wooden ladder or pole frame to be carried by hand like a stretcher or dragged behind a car or other vehicle. A hand cart for carrying a quarter-ton load could be built easily using old bicycle wheels. The problem with this simplest system is time, since most people with family cows will have to be making hay around the demands of other jobs. Most people, too, would need weeks just to learn the old art of mowing by hand.

To get around this first problem of getting the hay cut, almost everyone turns to some kind of mowing machine. There are several riding or walk-behind small tractors with attachments for cutting three- or four-foot swaths of long hay.

Our first machine was a horse-drawn riding mower with the hitch converted to a single tow bar. We towed with a four-wheel-drive car. It had the clearance for field work and the low speed that has to be the way with these rigs.

If these old mowers can be found they are usually very inexpensive. They work well behind a tractor or any low-speed vehicle but they can be dangerous on rough, rocky ground. If the cutter-bar snags, the machine lurches, and whoever is on the seat can get an awful send-off. It's best not to use chains or four-wheel drive on the tow vehicle because it is better to slip and skid a little than to pull the mower through a boulder.

KEEP THEM SHARP

The secret with these mowers and all other cutter-bar machines is to keep them sharp and run them slowly. A dull mower run fast crams with grass, snags on every rock and soon gets smashed. I've seen people mowing with oxen, so you know there isn't a too-slow speed for a machine that is tuned and sharp.

Cutter-bar mowers can be lethal. Cats often make the mistake of jumping into them, thinking, I guess, that they've stumbled on a host of mice. Keep an eye out for baby deer, too, bedded down and instinctively frozen in the tall grass.

Raking can be done by hand but again it is slow. Horse-drawn dump rakes still are common and cheap and are ideal for making a few acres of hay. Newer but old side-delivery rakes can be bought for $100 to $200.

For picking up loose hay the ladder stretcher for carrying a couple of hundred pounds at a time is OK close to the barn or winter stack. A ladder-like drag can be used to haul larger loads by car over fairly level and smooth ground.

With a little practice close to half a ton of hay can be carried on the back of a half-ton pickup truck. From there on, two- or four-wheeled wagons can be bought or built for carrying any load.

Along with a large wagon or truck might go a 1940s overhead loader that automatically picks hay from the windrow and feeds it up to one or two people standing on the building load.

The old buck rake that was fairly common 30 years ago would be an ideal tool for bringing in loose hay on a small place. Again, though, it would have to be a place with fairly level and rock-free ground. The buck rake is like a fork lift with four or more long, hardwood or steel tines. They can be mounted on the front or rear of a factory or home-built tractor in such a way that the tines can be raised a foot or so off the ground by winch or hydraulic system.

With any luck the hay is picked up by running the tines of the buck rake down a windrow. The hay stuffs itself on the tines—almost like baling—until no more will hold. Then you raise the rake and make for the barn. A lighter, finer hay may have to be stacked and tied on the tines.

Haying Weather

A day's mowing begins as soon as the dew is off the grass or it's dried from the last rain. On a good day—hot, sunny, and a light breeze blowing—mowed hay may be ready to rake within four or five hours. Or maybe it will take until the next noon before the mown grass on top of the swaths is beginning to crisp and is ready to turn under.

Raking by any means is a turning of the hay into long, snakelike windrows that catch the breeze. A side-delivery rake automatically swirls one swath at a time into these narrow rows. With a dump rake someone has to kick or trigger the mechanism that lifts the tines every 10 or 20 feet, building the windrows step by legs with each pass around the field. Although these old dump rakes were built to be operated by a driver riding on the rake itself, the dumping mechanism can be re-worked for tripping by a rope from the tow vehicle.

With good weather the hay in the first windrows may be ready for storage soon after the last swath is raked. On the other hand it may be necessary to turn the windrows a time or two with a pitchfork or the dump or side-delivery rake.

Sometimes it rains just as the hay is ready to turn into windrows. It can't be helped. The hay in the swath has to be turned by hand with a fork or by "tedder." Tedders are outstanding little machines that kick and fluff the hay, taking hours out of the normal time between swath and windrow. New tedders cost a lot. Old horse-drawn tedders should be snapped up before the flat-black antiquers figure what sort of ornaments to make out of them.

If the hay is baled, buck-raked or picked up by an overhead loader it can be taken directly from the windrow. I doubt that anyone making less than 20 tons of hay a year will want to buy a baler, but it might be possible and worthwhile to hire someone to bale up a few acres of hay.

Hay Cocks

If the hay is going to be fork-loaded onto a wagon or truck the windrows will want to be gathered or "tumbled" into small piles called *cocks*. In the days before balers, loaders and buck rakes, the hay was always cocked—first, because it was easier to handle this way, and second, because they wanted the hay to go through what they called the "sweat," a mild fermentation.

This sweating isn't necessary but it doesn't hurt. Certainly if hay isn't quite ready to go into final storage some evening or before a rain it is better to pile it up this way and let it sweat than to leave it open to a full wetting of dew or downpour.

A hay cock that is put together with some care will shed most any rain. When the sun shines again the surface dries fast, the cock is broken apart, and within a few hours the hay should be ready.

There probably are many favorite ways to build a proper hay cock. Whatever the system used, the final result should be steep-sided and peaked. Here is one system suggested by a Vermont farmer:

Start at one end of a windrow and fold 8 or 10 feet over and to the center of the hay that follows in the row. Then start down the line and bring about the same amount of windrow back and over the first. Gather up the "scatterings" for a cap on top of the cock. Maybe comb down the sides of the cock with the pitchfork and add these loose ends to the cap.

In humid climates or areas with heavy stands of legumes, it may be worth following a North European custom of gathering the hay against racks or tripods to help with air circulation.

Handling Hay

A bit of skill goes into handling and loading loose hay with a pitchfork. Some people make a mistake of using the fork as if it were a shovel with an underhand stab and throw that leaves a lot of hay behind. There are times for "shoveling" with a pitchfork; but it is a much more versatile tool that can be used for raking and sweeping and for latching on to 30 or 40 pounds of hay at a time for easy pitching to the top of an eight-foot load.

To grab a pile of hay, bring the fork tines under a leading edge, lift up and over as if you were forking up a piece of pancake. Stab down through the pile and lift. Raise the fork load overhead and back-hand it onto the load or wherever it's to go. If there is someone up there stacking or trampling the hay be careful. Pitchforks are also mean weapons.

Another way to get stuck is in buying a cheap pitchfork. Their handles are too short, their tines are brittle, short or have the wrong curve so that they are forever jabbing into the ground. A more expensive fork is probably worth the money.

In loading hay on a truck, a wagon or drag, start at the corners (usually the rear corners) and work forkful by forkful up each side with each pile overlapping the last in the line. Scatterings or small forks of hay go to the middle of the load until a definite valley up the center calls for a tie-in batch of the larger forks of hay. If no one is on the load someone will have to climb up now and then to tramp the hay down.

After working back and forth, building up as many tiers as seems reasonable, overlap three or four good forks of hay down the middle to complete the tie. It's amazing how much loose hay will go on a small wagon or truck if it is well tramped and lapped together.

More loose hay can be carried on a small truck if the tailgate is

The overhand toss pitches loose hay to the top of the load in this photograph taken around the turn of the century. Although baled hay is much more common on commercial farms today it's likely that many family cow owners will find it as easy and cheaper to handle and store the loose variety. (Photo courtesy of Vermont Extension Service archives.)

left down. A couple of sturdy posts at the back corners of the bed are better than side boards. Let the hay hang over the sides and back. It will all stay on if it's lapped into the load.

In moving the hay from the truck or wagon to a mow, try to remember where the last fork went on. Start there and work back, fork by fork to the first in the load. Otherwise there can be an awful lot of wasted effort trying to tear the load apart.

The best mow for loose hay is a second-floor loft with a pole floor that lets air up through the mass. In the mow, go up by tiers, spreading each new load of hay over as wide an area as possible. If the hay is still a bit damp, spread it out and don't tramp it down. Leave the barn doors and windows open for all the ventilation possible. Remember how much more weight there will be in densely-packed bales of hay and don't overload an old hay loft.

Don't pile hay on a damp earth floor. An outside stack should be built on top of a rough stage or crossed poles and should be at least capped with a tarpaulin well anchored with ropes or old tires.

Testing Hay for Storing

It's hard to explain when the hay is right for storing. With loose hay, I go on color, the feel and the sound of the hay. I probably tend toward the too-dry and heavy-shattering side of things. I've never bothered to explain why, but come to think of it I can better imagine buying protein and vitamin supplements to make up for a poor-quality hay than I can the thought of having to build a new barn because of fire.

The hay I put away is green, but not the livid green of new-mown grass. It's more a dusky blue-green. It doesn't feel damp but it is still pliable enough that a hank could be twisted and tied in a knot. When I kick it, ready hay talks back with a whisper and rustle. Green hay moves quietly aside like a tired old dog.

DEXTER'S SALT METHOD

A much more precise and yet simple method for finding out when hay or grains are dry enough for storage has been worked out by Professor S. T. Dexter at Michigan State University. Samples are shaken with salt in a closed and dry container and the time it takes before the salt begins to clump comes very close to telling how much moisture is in the hay or grain.

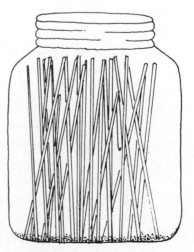

Hay sample being given the Dexter hay-moisture test.

To check hay this way, get a wide-mouthed quart jar and dry the air inside it thoroughly by putting in a spoonful of salt, capping and shaking it around for a couple of minutes. This salt and the salt for testing the moisture in the hay has to be one that hasn't been doctored with chemicals meant to keep it from taking up water. Prof. Dexter suggests "any cheese salt, Diamond Crystal Shaker or Diamond Butter Flake salt"—but *not* the Diamond Crystal Weather-Pruf variety.

Go around the field gathering hay from under or around the rows or piles of raked hay. From this overall sample take a good handful, twist it to split the stems, and cut off both ends leaving a middle portion that will fit comfortably inside the quart jar. Uncap the bottle, dump out the old salt, put in the hay and add a fresh teaspoonful of free-running salt. Cap and shake about 100 times.

Now tap the bottle to bring the salt to one end. If the salt is already clumping, the hay is definitely too wet for safe storage. If the hay is borderline, it will be one to two minutes before the salt shows signs of picking up moisture. If after two minutes the salt is still free-flowing the hay can be stored loose. Hay that is to be stored baled or chopped should go at least five minutes before the salt begins to show the moisture.

WHEAT AND GRAINS TESTING

For checking wheat, oats, or barley, Prof. Dexter uses a smaller jar, two to four ounces, and he covers it with a paper collar to keep out light and hand warmth. The jar is salt-dried, emptied, and about half filled with a random sample of the grains to be checked. Add about a half teaspoon of dry, free-flowing salt, cap and shake for two to three minutes. Let stand to a total of five minutes. If then the salt is still "perfectly free-flowing" the grain may be stored safely. (The recommended safe moisture levels for storing these grains is 14 to 15 percent.)

Spontaneous Combustion

The most critical time for high-heat spoilage and fire is over the first two months of storage. If you are worried, take the temperature of the mow. Use any thermometer that reads up to 200° F. (93° C.).

You want the temperature deep in the hay, so either drive a pipe down and lower the thermometer on a string or tape the thermometer to a fork handle and push this down. Take the temperature here and there around the mow. Keep a nose out for smells of burning.

If the temperature in the depths of the mow is up around 140° F. (60° C.) there is a possibility that trouble is brewing. Wait a few hours and try it again. If the mow temperature has gone up while the air in the barn has stayed constant or has dropped, take the mow apart.

If the temperature goes to 160° F. (71° C.) or above, there is a definite danger of fire. Flames may only be waiting for a rush of fresh air. In this case *don't take the mow apart until you have a fire department or plenty of fire-fighting equipment behind you.*

Years ago it was common practice to throw some rock salt around on top of loose hay to prevent spoilage and fires. I've heard people say it works, but Prof. Morrison says it isn't guaranteed to prevent either, though it may make a poor-quality hay more palatable.

Better than using salt is to put the hay away fully cured. Spread it out. Don't go crazy tramping it down—just enough to get what hay you need into the barn. And leave the barn doors open through dry days during haying and on into summer.

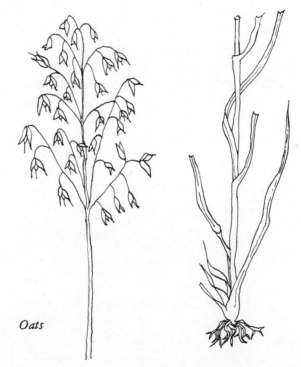

Oats

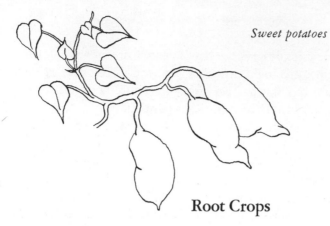

Sweet potatoes

Root Crops

Forage crops are more easily grown than roots. However, you can harvest and store roots more easily than you can make and store hay by the medium-oldfashioned way I've described.

Roots, of course, are an annual. They have to be planted in soil that is as well prepared as a vegetable garden, or nearly so. Most of them grow best in a climate that gives them 10 to 15 inches of rain spread more or less evenly over the season. People living in drier areas could irrigate the crop, but then again they might better think about growing corn or one of the sorghums for fodder.

For quarter-acre or smaller plantings of beets, turnips, and mangels, a homemade drill (seeder) can be made out of a jar taped upside down to the end of a cane or stick. Punch a small hole in the cap. Fill the bottle with seeds and plant by walking down the rows giving the bottle a tap against the ground every foot or so. Three or four seeds should sprinkle out with each tap. This little drill saves your back while planting and saves hours when the time comes to thin the rows.

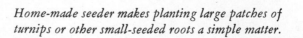

Home-made seeder makes planting large patches of turnips or other small-seeded roots a simple matter.

236

ROOT STORAGE

Roots other than potatoes may store well enough in the ground in areas with mild climates. We have kept turnips in the ground where the temperature sometimes dropped to zero. They kept better when covered with a thick layer of old hay. If roots can't be left in the ground they have to be sliced and dried or else stored away whole in a cool, damp (but not wet) cellar.

If the barn hasn't got a root cellar, build one into the side of a hill or wherever the drainage is best. The sides of the cellar or the container for the roots may be stone, rough boards or poles, leaving cracks for ventilation. A separate root cellar should have a capped vent pipe to let moisture escape. It should have a framed door or at least have something stacked against the opening like old feed bags stuffed with hay or sawdust.

From Scotland to France roots often are stored in above-ground "clamps." These are shallow earth pits bedded with straw or old hay, filled with roots and covered with more straw and a layer of dirt. For

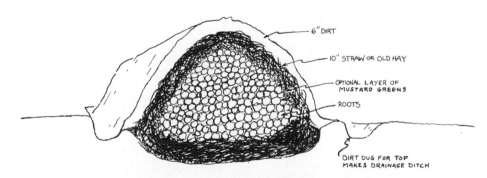

6" DIRT

10" STRAW OR OLD HAY

OPTIONAL LAYER OF MUSTARD GREENS

ROOTS

DIRT DUG FOR TOP MAKES DRAINAGE DITCH

"Clamp" for outdoor root storage. This system or variations used in England and Central Europe could be used by people who haven't a barn basement or root cellar for storing root crops. The thickness of the insulating layers of dirt and straw or old hay would have to be varied depending on the severity of the winter climate. Years ago some farmers would place a layer of mustard or other greens over the roots before covering them with straw and dirt. Presumably methane produced as the greens decomposed helped preserve the roots.

ease of feeding they are usually built long and narrow. The roots are stacked to a peak.

In Roman times these piles of roots sometimes were covered with a layer of wild mustard plants before the straw or hay and dirt went over the top. The mustard greens decomposed, producing methane which apparently helped preserve the stored crop.

One More Thing

There's enough for now. Wealth is a barn full of hay, a few hundred pounds of grain and a cow giving down her milk while cats weave restlessly, ready to spring to the dish by the door. That reminds me: Don't bother to teach cats how to catch a squirt direct from the spigot. They'll never leave you alone then. I had one cat worked into a routine where she could get a squirt if she meowed, licked her paw and bit my left arm—in that order. I had a damned sore arm before she unlearned her lesson. It served me right.

It's more fun seeing if I can squirt my son Wim when he peeks over the side of the shavings bin. There's a tough angle on that shot, but I got him the other night—got him good—and he laughed.

Tonight it's snowing, and white triangles are pointing up the black cold beyond each window pane. Gusts of wind shearing past the north gable make the rafters creak, like winter shifting in her sleep. But the old barn's alive again within.

A pig grunts, grudging another's right to push in where the pine shavings are deep and dry. How many pigs and cats and calves has this old barn seen in her years—and women and men and boys or girls sitting here in this same corner milking? Some of their names are carved on the faces of beams and doors around the barn. Some of the same and other names are carved on stones, erect and fallen, up there on the hill behind the house.

I wonder if any of them were "into" cows. Was it all a chore? Or were there lucky people like me who for unexplainable reasons found that raising a few animals, a garden of vegetables and especially a cow, was somehow "into" them?

Glossary

aftermath: pasturable forage growth that follows the harvesting of a grain or hay crop.

bag: udder.

bagging (or to bag up): the rapid expansion of the udder that may begin up to four weeks prior to calving.

balanced ration: a daily ration of food that fully satisfies a cow's nutrient requirements.

beef: the meat of a cow, steer or bull.

beef (verb): to slaughter a cow, steer or bull for its meat as in, "They plan to beef the cow soon."

beef animal: any cow, steer or bull destined to become or used for the production of meat. Usually the term used for animals belonging to the so-called beef breeds.

beef type: any cow, steer or bull showing the stocky, fleshier form that is favored by people concerned with meat production.

bloat: severe distention of the rumen through the build-up of trapped gases given off in the processes of fermentation.

bob calf: any newborn calf under a week old. Usually weighs 60 to 100 lbs.

bob veal: the flesh of bob calf.

bulk: the amount of physical space taken up by a food in relation to the nutrients it contains. Hay and other bulky foods take up much space relative to the nutrients they provide. Concentrated foods such as grains or oil meals lack bulk.

bull: male bovine.

bulling: a cow in heat or estrous is sometimes said to be bulling, perhaps because in ways her behavior may become bull-like. For instance, she may become nervous, bellow repeatedly and mount other cows as though she were a bull.

bullock: usually a castrated bull though it may be used to refer to a young bull.

calf: any young cow or bull. Although the term is usually used for animals under about ten months of age there is no sharp line. Males and females are either bull or heifer calves.

cereal: any plant of the grass family that yields an edible, starchy grain. Corn, wheat, rye, oat, barley, rice etc.

choke: a blockage of the gullet (esophagus), frequently caused by a potato or other small root or fruit swallowed whole. Subsequent bloat may complicate the problem.

cleanings: the afterbirth (placenta) expelled following calving.

cock: a small pile of hay.

colostrum: the high-protein, vitamin and antibody-rich "first milk" produced by a cow at the time of calving.

concentrate: a food that is high in energy and/or protein value and, in most cases, lacking in bulk.

condensed milk: an evaporated milk sweetened with sugar.

cooling: a word once commonly used to describe foods that are relatively low in energy value. Among the grains whole oats would be said to have a cooling effect on the animal being fed and the reason could be their relatively high content of slowly-digestible fiber. (See "heating," below.)

cow: any mature female bovine, be she beef, dairy or whatever type.

cream line: the visible line that develops in time between cream and milk when whole, unhomogenized milk sits quietly, so that the lighter-than-water butterfat globules are able to rise.

crossbreeding: the breeding of a dam belonging to one pure breed to a sire from another. Crossbreeding may be carried out in attempts to combine characteristics of parent breeds in a new breed as was done in the development of Brangus cattle (out of Brahmans and Angus), or for the creation of offspring exhibiting what is known as hybrid vigor (heterosis).

cutting, second (third, fourth and on): follow-up cuttings of hay taken off a field at six- or eight-week intervals through a growing season. See "rowen" below.

dairy type: any cow or bull—though more in reference to cows—having the body form preferred by those raising animals for milk production. In cows the dairy type is characterized by a relative fineness of line and a wedge-like profile.

dam: a female parent.

deacon: a term sometimes used for a bob calf.

dent: a variety of corn, *Zea mays indentata,* whose kernels become dented in the last stages of maturation and drying.

dough: one of the later stages in the development of a cereal grain. Grains in the dough stage are soft but not juicy as they are when they are in what is known as their *milk stage.*

drop: another word for gutter (see below).

dry cow: any cow that is not giving milk. It may be said that the cow is not "in milk" whereas a cow that is giving milk is said to be "in milk" or "in lactation."

dugs: teats.

evaporated milk: unsweetened concentrated milk. The first patent for a concentrated milk was granted in 1856 to Gail Borden who developed his process in a laboratory provided by the Shaker religious community at New Lebanon, New York.

flint: a variety of corn, *Zea mays indurata,* whose kernels are smooth and rounded in the last stages of maturation and drying. Flint corn has more hard, flint-like starch grains than has the dent type. It is often preferred over dent corn in northern states and Canada because it more readily germinates in cool soils.

fodder: the seed head, stalk and leaves of a grain such as corn or sorghum harvested entire.

forage: pasture and field crops that are of value as producers of edible leaves and stems.

formula feed: a mixture of concentrated foods mixed in the proportions necessary to balance a cow's ration.

freemartin: a sexually imperfect and usually sterile heifer calf born twin to a bull calf (see Chapter 15).

girthing tape (or chain): a measuring tape or fine-linked chain used to translate the diameter of a cow or steer's heart girth into an estimate of the animal's live weight.

goad: a stick or other instrument used to drive cattle. A stout wooden cane is used as a goad on many farms and is to livestock auctioneers what the gavel is to a courtroom judge.

grade: any cow or bull not registered with one of the pure breed associations.

grading up: the practice of improving "grade" stock (stock of uncertain quality and pedigree, not registered stock) through the breeding of females to sires of the type and breed desired.

grain: the seed of any member of the grass family (themselves often referred to as grains) that is rich in easily digested starches.

grass: plants of the scientific family name, *Gramineae*. They have long leaves with parallel veins, jointed stems and bear flowers and seeds (sometimes classified as grains) at the ends of central stalks.

gutter: the shallow trench that collects the manure behind a cow's tie or stanchion stall. See "drop," above.

heat, estrous: the period of ovulation and readiness for breeding and fertilization—24 to 36 hours altogether—that repeats itself on an average of every 21 days in sexually mature cows.

heating: a word once commonly used to describe a food that is relatively high in energy value. Among the grains, corn would be a *heating* or *hot* food because it is rich in easily-digested starches and low in fiber.

heaving: a winter damage to plants, particularly to those like alfalfa that are deep-rooted, that comes about when the soil surface expands on freezing, heaving upward and so pulling the plant roots apart.

heavy soil: a soil having a high proportion of clay particles, the smallest mineral soil particles after sand and silt. Heavy soils may be difficult to till. The tightly-packed clay particles often inhibit the movement of water, nutrients and plant roots through the soil.

heifer: a young beef or dairy cow. Usually one that has not yet had a calf, though sometimes a young cow that has calved may be called a "first-calf heifer."

inbreeding: the breeding of closely related animals, such as sire to daughter or daughter to son. Inbred offspring will reflect a magnification of the worst traits in both parents. Inbreeding may be engineered by experienced livestock breeders in attempts to reinforce or enhance good characteristics seen in the parents. The practice should not be attempted by the inexperienced because there is too great a likelihood that bad heritable traits will be magnified in the offspring.

lactation: the physiological processes of milk production.

leaching: the downward percolation of minerals and organic nutrients through the soil, often beyond the reach of plant roots.

lead: the long end of a halter or neck rope by which a cow is tied or led.

legumes: plants that are members of the family *Leguminosae*. Sometimes called tri-foliates because their leaves come in threes, the legumes have root nodules harboring nitrogen-fixing bacteria. Beans, peas, alfalfa, clovers, vetches, soybeans and peanuts are included in this large family.

light soil: a soil that is porous and loose due to its having a high proportion of sand. Though often preferable to heavy soils so far as crop production is concerned, light soils may tend to dry out rapidly and lose nutrients through leaching.

linebreeding: a form of inbreeding in which offspring are bred back to one common ancestor.

loam (pronounced *lome* or *loom*): a type of soil that is rich and best suited to tillage and crop production due to its ideal mixture of sand, silt, clay and organic matter.

lodge: to collapse as happens to grain crops especially when they are heavy with seeds. High winds, heavy rain or snow may cause severe lodging in crops nearly ready to harvest. Badly lodged grains may be all but impossible to harvest by machine.

mastitis: inflamation of the udder frequently caused or severely aggravated by an invasion of bacteria.

milch cow: an old term for the milk cow.

milk: a pre-dough stage in the development of grains. A grain in the milk stage may be fully formed but soft and juicy when crushed.

near side: in most cases this is the cow's or ox's left side (looking at the animal from behind). Customarily a cow or ox is led from this side. In the case of a team of oxen the one on the left is called the *near ox*.

off side: in most cases the cow's or ox's right side. In a team of oxen the one on the right is called the off ox.

open: a cow or heifer of breeding age that has not been bred.

ox, oxen: an adult, castrated bull of any breed that has been trained for work. Historically also a term for any male bovine.

poke: a device usually made from wooden boards or poles in the form of a letter H or A and placed around the neck of a cow, calf or whatever, to keep the animal from pushing through fences.

poll: the top or crown of a cow's head between the horns. In the case of polled (hornless) animals or cows whose horns were removed the poll stands out as a prominent ridge.

polled: a genetically hornless cow or bull.

pure bred: a cow or bull born of parents belonging to a recognized breed.

ration: a cow's daily intake of all foods.

registered: a cow or bull certified as being pure bred through the act of registration with one of the breed associations.

roughage: a food that is relatively high in fiber and bulk, and low in digestible nutrients. The opposite of a concentrate.

rowen: second cutting or aftermath.

scours: a term used for diarrhea, particularly in calves.

settle: conceive. A cow that fails to conceive has not settled.

sheaf (sheaves): a bound bundle of grain stalks, entire.

sire: a male parent.

shock (sometimes shook or stook): a stand of four or more sheaves left in the field for further drying.

short feed: a term used in Canada and perhaps elsewhere for concentrate mixtures or formula feeds.

silage (ensilage): a green roughage preserved through fermentation in upright silos or trenches covered with plastic tarpaulins.

silo: a pit, trench or upright tower (usually cylindrical) used for the storage of silage.

skimmed milk: milk from which most of the butterfat has been removed.

snapped (corn or sorghum): the entire grain head with husk.

soilage: a crop, usually an annual, that is cut and fed green.

solids: the non-water ingredients of milk which include fat, proteins and minerals. Together the solids add up to a maximum of about 15 percent of the richest whole milk.

solids-not-fat: the total solids in milk minus the butterfat portion.

sour (soil): a soil that is distinctly acid in reaction.

springer: a cow that is about to calve.

stanchion: an elongated wooden or metal frame that clamps loosely behind a cow's head confining her to a stall, while at the same time allowing her freedom to move back and forth a foot or two and to lie down.

standing heat: that time, beginning some hours after a cow has entered a heat period and lasting up to 12 hours, when the cow will stand still while being mounted by other cows or a bull.

stary: the dry, rough and uneven-coated condition seen in animals that are diseased or malnourished.

steer: a bull castrated while still a calf and being raised for beef.

stover: the stalks and leaves of a grain with the seed head removed. Corn and sorghum stovers are used as low-quality roughages.

succulent: a general term for wet roughages and roots.

sucker: weaning rings. Sometimes called *calf suckers*. (See *weaner*.)

supplement: any addition to an otherwise balanced ration.

sweet (soil): a soil that is slightly acid to slightly basic (alkaline) in reaction.

teat: the elongated spigots on a cow's udder. May be pronounced *tit* or *teet*.

test: the measure of butterfat in milk. The first reliable butterfat test was developed by Stephen Babcock in 1890.

udder: the body of the cow's mammary gland, divided internally into four separate quarters, each of which is normally equipped with one teat.

veal: the pale meat of a milk-fed or milk-substitute-fed calf up to about three months of age. The meat lacks color primarily because the calf is kept on a diet that is lacking in iron.

vealer: a bob or older calf destined to become veal.

weaner: any device with prongs or sharp points fitted to or over the nose of a calf to keep it from nursing.

yoke: a wooden frame around the neck or tied to the horns of an ox. A pole or poles are attached to the yoke for pulling a cart, drag or field implements. "Yoke" may also be used when referring to teamed pairs of oxen.

Other Books and Places

New Books

Dairy Cattle: Principles, Practices, Problems and Profits, by R. C. Foley, D. L. Bath, F. N. Dickinson and H. A. Turner. Lea and Febiger Publishing Co., Philadelphia, Pa.

Feeds and Feeding, Abridged, by Frank B. Morrison. The Morrison Publishing Company, Claremont, Ontario, Canada. (Adapted from *Feeds and Feeding,* 22d edition.)

Fences, Gates & Bridges, edited by George A. Martin. Stephen Greene Press, Brattleboro, Vt.

Merck Veterinary Manual. Merck and Company, Rahway, N.J.

Part Time Livestock Farming, by James G. Welch, Ph.D. Available through author: c/o Dept. of Animal Sciences, University of Vermont, Burlington, Vt.

Raising Milk Goats the Modern Way, by Jerry Belanger. Garden Way Publishing, Charlotte, Vt.

The Butchering, Processing and Preservation of Meat, by Frank G. Ashbrook. Van Nostrand Reinhold Co.

Veterinary Guide for Farmers, by G. W. Stamm. Popular Mechanics Press.

Publications

"Dairy Husbandry in Canada," Publication No. 1439, Information Division, Agriculture Canada, Ottawa. K1A 0C7

U.S.D.A. List No. 11, "List of Available Publications" (45 cents), Superintendent of Documents, U.S. Government Printing Office, Washington, D.C. 20402

"Nutrient Requirements of Dairy Cattle" (#ISBN 0-309-01861-7).
"U.S.-Canadian Feed Composition Tables" (#ISBN 0-309-01684-3).
  Both from Printing and Publishing Office, National Academy of Sciences, 2101 Constitution Ave., Washington, D.C., or National Research Council of Canada, Ottawa, Ont., Canada.

Raising Veal Calves, Publication No. 106 of the Massachusetts Extension Service, University of Massachusetts, Amherst, Mass.

General Publications

Local Agricultural Extension Service representatives may be found in the phone book Yellow Pages. Through these offices come addresses for provincial, state or federal agricultural institutions, printing offices and testing facilities. Get on their mailing list for regular news letters.

Old Books, Out of Print

Worth looking for in old-book stores:

  Soils and Men. U.S.D.A. Yearbook of Agriculture, 1938.

  Food and Life. U.S.D.A. Yearbook of Agriculture, 1939.

  Feeds and Feeding (complete or abridged), by Henry and Morrison. (The first *Feeds and Feeding,* forerunner of the present book by Frank B. Morrison, came out in 1898. This and later editions through the 1920s devote more space to feeds and management techniques that aren't as important to commercial farming today but which may be helpful to part-time farmers.)

Things

Butter Churns:

  Sears, Roebuck and Company, Boston, Mass. (electric churn).

  Gem Dandy Electric Churns, Alabama Manufacturing Co., 2110 1st Ave. N., Irondale, Ala. 35210.

Ice Cream Freezers:

  White Mountain Freezer Co., Winchester, Mass.

Cream Separators:

  Centrifugal cream separators (hand and electrically operated), European models like DeLaval and others, also parts. North American importer

George Heidebrecht, Foothills Creamery, 4207 16th St., Calgary, Alberta, Canada.

Associations

Registered Dairy Cattle:

The American Dairy Cattle Club. Robert W. Hitchcock, Interlaken, N.Y.

The American Guernsey Cattle Club. 70 Main Street, Peterborough, N.H.

The American Jersey Cattle Club. 2105J South Hamilton Rd., Columbus, Ohio 43227.

The American Red Danish Cattle Association. Marlette, Mich.

The Ayrshire Breeders' Association. Box 1038, Beloit, Wis. 53511.

The Dutch Belted Cattle Association of America. Miami, Florida.

The Holstein-Friesian Association of America. Box 808, Brattleboro, Vt. 05301.

Red and White Dairy Cattle Association. Box 771, Elgin, Ill. 60120.

The Canadian Cattle Breeders' Society (Société des Eleveurs de Bovins Canadiens). Roxton Pond (Shefford), Quebec JOE 1ZO, Canada.

The Purebred Dairy Cattle Association. 70 Main St., Peterborough, N.H.

Ayrshire Breeders' Association of Canada, 1160 Carling Ave., Ottawa, Ontario, K1Z 7K6.

Canadian Brown Swiss Association, 358 Wellington St., St. Thomas, Ontario, N5R 2T5.

Canadian Jersey Cattle Club, 343 Waterloo Ave., Guelph, Ontario, N1M 3K1.

Canadian Guernsey Breeders' Association, 368 Woolrich St., Guelph, Ontario.

Holstein-Friesian Association of Canada, Brantford, Ontario.

Dual-Purpose Cattle:

The American Milking Shorthorn Society. 313 South Glenstone Ave., Springfield, Mo.

Red Poll Cattle Club of America. 3275 Holdrege St., Lincoln, Nebr.

Canadian Shorthorn Association, 5 Douglas Street, Guelph, Ontario, N1H 2S8.

Canadian Red Poll Cattle Association, Francis, Saskatchewan, SOG 1VO.

Appendix

**AVAILABLE CALCIUM AND
PHOSPHORUS SOURCES**

Supplement	Calcium (%)	Phosphorus (%)
Calcium phosphate*	15.5–18.5	18.5–21.0
Calcium carbonate ($CaCO_3$)	38.0	00.0
Limestone, ground	33.0	00.0
Oyster shells	33.0–38.0	00.0
Tricalcium phosphate	38.0	18.0
Monocalcium phosphate	20.0	21.0
Defluorinated phosphate	31.0–34.0	18.0
Dicalcium phosphate	26.0	18.0
Disodium phosphate	00.0	21.6
Steamed bone meal	24.0–28.0	12.0–14.0
Monosodium phosphate	00.0	25.0
Sodium tripoly phosphate	00.0	25.6

---

* Various commercial products are available which are chemical mixtures of dicalcium and monocalcium phosphates; these usually are referred to as dicalcium phosphates. The actual percentage of calcium or phosphorus will vary from one product to another.

(From *Extension Bulletins*, Florida, Atlantic Provinces.)

FAMILY COW FEEDING STANDARD

Daily Nutritional Requirements for Mature Cows

(1) Maintenance

		DRY MATTER INTAKE					
	Cow Weight, (lbs.)	*Average Hay (lbs.)*	*Maximum (lbs.)*	*Crude Protein, (lbs.)*	*TDN (lbs.)*	*Ca (lbs.)*	*P (lbs.)*
Jersey	800	16	28	1	6	0.07	0.04
Guernsey							
Ayrshire	1000	20	35	1.1	7	0.09	0.05
Holstein							
Brown Swiss	1200	24	42	1.3	8	0.10	0.05

(2) Milk Production

	BUTTERFAT							
	HOLSTEIN	BROWN SWISS	AYRSHIRE		GUERNSEY		JERSEY	
	3%		*4%*		*5%*		*6%*	
Pounds	*CP (lbs.)*	*TDN (lbs.)*	*CP (lbs.)*	*TDN (lbs.)*	*CP (lbs.)*	*TDN (lbs.)*	*CP (lbs.)*	*TDN (lbs.)*
20	1.0	5.5	1.5	6.5	1.5	7.5	1.5	8.5
30	2.0	7.0	2.0	12.5	2.0	11.0	2.5	12.5
40	2.5	11.0	3.0	9.5	3.0	14.5	3.3	16.5

Pounds	*Ca (lbs.)*	*P (lbs.)*
20	0.08	0.06
30	0.12	0.09
40	0.16	0.12

FEED ANALYSIS TABLE (As-Fed Basis*)

	Dry Matter %	Crude Protein %	TDN %	Fiber %	Ca %	P %
Dry Roughages						
Alfalfa	90	15.5	51	29	1.5	0.2
Bromegrass (bloom)	90	8.5	49	28	0.4	0.2
Mixed hay, good,						
< 30% legumes	90	9.0	49	31	1.0	0.2
Oat Straw	90	4.0	45	36	0.24	0.1
Orchard grass	90	8.0	49	30	0.25	0.2
Prairie hay						
(midseason)	90	6.0	45	—	0.3	0.1
Quack grass	90	7.0	40	35	—	—
Timothy, good	90	6.5	49	30	0.35	0.14
Timothy, late	90	5.5	41	31	0.14	0.15
Timothy & ¼						
clover	90	8.0	50	30	0.58	0.15
Corn fodder	80	7.0	55	21	0.25	0.14
Corn stover	80	6.0	45	27	0.48	0.08
Succulents						
Apples, whole	18	0.5	13	1.5	0.01	0.01
Apple pomace	21	1.5	16	4.0	0.02	0.02
Cabbage	9.5	2.0	8	1.0	0.06	0.03
Carrots	12	1.0	10.5	1.0	0.05	0.04
Corn silage	27	2.5	18.5	7.0	0.1	0.07
Kale	12	2.5	8	1.5	0.2	0.06
Mangels	10	1.5	7	1.0	0.02	0.02
Potatoes	21	2.0	17.5	0.5	0.01	0.05
Rutabaga (Swede)	11	1.5	9.5	1.5	0.05	0.03
Sorghum silage	25.5	1.5	15	7.0	0.08	0.05
Sweet potatoes	31	1.5	25.5	2.0	0.03	0.04
Pumpkins	10	2.0	9	2.0	—	—
Sugar Beet (tops)	16	1.5	14	1.0	0.03	0.07
Turnips	10	1.0	8	2.0	0.06	0.02

---

* Meaning that in order to calculate nutrient values of these feeds you simply multiply the percentage CP or TDN times the total weight of food being fed or being consumed. In other tables you may find CP and TDN values given under the heading *Dry Matter Basis*. In that case it is necessary first to find out the dry matter content of the weight of the feed being fed or consumed before going on to calculate your CP, TDN or other values.

	Dry Matter %	Crude Protein %	TDN %	Fiber %	Ca %	P %
Grains						
Barley	89	13	77	5.5	0.06	0.35
Corn	88	9	83	2.0	0.02	0.30
Oats (whole)	90	12	70	12.0	0.09	0.30
Sorghum	89	11	80	2.0	0.05	0.30
Pasture						
Mixed grasses	20	4	15	6.0	0.60	0.30
Protein Supplements						
Cottonseed meal	91	41	70	11	0.15	1.1
Corn gluten meal	91	43	80	4	0.16	0.40
Linseed oil meal	91	35	73	9	0.40	0.80
Wheat bran	89	16	66	10	0.10	1.2
Wheat middlings	90	18	78	4	0.10	0.80
Commercial concentrate, premixed	90	12–22	70–75	11	0.60	0.50

RATES OF APPLICATION EQUIVALENTS
(U.S. MEASURES)

1 ounce per square foot = 2,722.5 pounds per acre.
1 ounce per 100 square feet = 27.2 pounds per acre.
1 ounce per square yard = 302.5 pounds per acre.
1 pound per 100 square feet = 435.6 pounds per acre.
1 pound per 1,000 square feet = 43.56 pounds per acre.
1 pound per acre   = 1/3 ounce per 1,000 square feet.*
5 gallons per acre   = 1 pint per 1,000 square feet.*
100 gallons per acre = 2.5 gallons per 1,000 square feet.*
100 gallons per acre = 1 quart per 100 square feet.*
100 gallons per acre = 18.5 pounds per 1,000 square feet.*

* Approximate.

CONCENTRATE FEEDING SCHEDULE

RECOMMENDED FEEDING OF CONCENTRATES (LBS. DAILY) FOR COWS ON EXCELLENT PASTURE (12–14% CP)

Milk produced daily (lbs.)	Butterfat in milk			
	3.0% (lbs.)	3.5% (lbs.)	4.0% (lbs.)	5.0% (lbs.)
30	—	—	1.0	1.5
35	1.5	2.0	2.5	2.5
40	2.5	3.5	4.5	5.0
45	4.5	5.5	6.5	7.5
50	7.0	8.0	8.5	10.0
55	9.5	10.5	11.0	12.5
60	12.0	13.0	14.0	16.0
65	14.0	15.5	17.0	19.5
70	16.5	18.0	20.0	23.0
E.F.O.* 75	19.0	21.0	23.0	26.5
80	22.0	24.0	26.0	30.0
85	25.0	27.0	29.0	—
90	28.0	30.0	33.0	—

RECOMMENDED FEEDING OF CONCENTRATES (LBS. DAILY) FOR COWS ON EXCELLENT HAY OR SILAGE, OR GOOD PASTURE (14–16% CP)

Milk produced daily (lbs.)	Butterfat in milk			
	3.0% (lbs.)	3.5% (lbs.)	4.0% (lbs.)	5.0% (lbs.)
15	—	—	1.0	1.5
20	1.0	1.5	2.0	3.0
25	2.0	2.5	3.0	4.0
30	3.5	4.0	4.5	5.5
35	5.0	5.5	6.0	7.5
40	7.0	7.5	8.5	10.0
45	9.5	10.0	11.0	12.5
50	11.5	12.5	13.5	15.5
55	14.0	15.0	16.5	19.0
60	16.5	18.0	19.5	22.5
E.F.O.* 65	19.0	21.0	22.5	26.0
70	22.0	24.0	26.0	30.0
75	25.0	27.0	29.5	—
80	28.0	30.0	33.0	—

RECOMMENDED FEEDING OF CONCENTRATES (LBS. DAILY) FOR COWS
ON GOOD HAY OR SILAGE, OR FAIR PASTURE (16–18% CP)

		Butterfat in milk		
Milk produced daily (lbs.)	3.0% (lbs.)	3.5% (lbs.)	4.0% (lbs.)	5.0% (lbs.)
15	—	—	1.0	1.5
20	1.0	2.0	2.5	4.0
25	4.0	4.5	5.0	6.0
30	6.5	7.0	7.5	8.5
35	8.0	9.0	9.5	11.0
40	10.0	11.0	12.0	14.0
45	12.5	13.5	15.0	17.5
50	15.0	16.5	18.0	21.0
55	18.0	19.5	21.0	24.0
60	21.0	22.5	24.5	28.0
E.F.O.* 65	24.0	26.0	28.0	32.0
70	27.0	30.0	31.5	—
75	30.0	—	—	—

* Recommended that Experienced Farmers Only feed these levels of concentrates.

(Adapted from *Dairy Husbandry in Canada,* Publication #1439, 1971.)

APPROXIMATE WEIGHTS OF COMMON
VOLUMES OF COMMERCIAL CONCENTRATES

Volume	Weight in Concentrates
1 pound coffee can	1½ lbs.
2 pound coffee can	2½ lbs.
48–50 fl. oz. juice can	2 lbs.

ESTIMATION OF WEIGHT ACCORDING TO HEART GIRTH

Circ. of Chest (in.)	Wt. in lbs.	Circ. of Chest (in.)	Wt. in lbs.	Circ. of Chest (in.)	Wt. in lbs.
30.0	100	51.0	414	72.0	1064
30.5	103	51.5	424	72.5	1085
31.0	107	52.0	434	73.0	1104
31.5	112	52.5	445	73.5	1126
32.0	117	53.0	456	74.0	1146
32.5	121	53.5	467	74.5	1169
33.0	127	54.0	476	75.0	1191
33.5	131	54.5	495	75.5	1213
34.0	137	55.0	510	76.0	1236
34.5	141	55.5	521	76.5	1263
35.0	146	56.0	534	77.0	1285
35.5	152	56.5	545	77.5	1308
36.0	157	57.0	562	78.0	1331
36.5	162	57.5	577	78.5	1354
37.0	167	58.0	590	79.0	1377
37.5	173	58.5	605	79.5	1400
38.0	179	59.0	616	80.0	1423
38.5	186	59.5	629	80.5	1446
39.0	193	60.0	647	81.0	1469
39.5	199	60.5	668	81.5	1492
40.0	206	61.0	684	82.0	1515
40.5	214	61.5	700	82.5	1538
41.0	223	62.0	716	83.0	1561
41.5	229	62.5	732	83.5	1584
42.0	239	63.0	748	84.0	1607
42.5	247	63.5	762	84.5	1629
43.0	255	64.0	778	85.0	1650
43.5	270	64.5	790	85.5	1673
44.0	283	65.0	815	86.0	1692
44.5	293	65.5	828	86.5	1718
45.0	298	66.0	848	87.0	1741
45.5	307	66.5	866	87.5	1764
46.0	315	67.0	883	88.0	1788
46.5	323	67.5	891	88.5	1812
47.0	334	68.0	904	89.0	1833
47.5	344	68.5	923	89.5	1857
48.0	354	69.0	942	90.0	1881
48.5	364	69.5	962	90.5	1905
49.0	374	70.0	982	91.0	1929
49.5	384	70.5	1002	91.5	1952
50.0	394	71.0	1022	92.0	1975
50.5	404	71.5	1043		

RULES FOR ESTIMATING GRAINS
AND ROUGHAGES

1. *To find the number of bushels of grain or shelled corn in a bin:* Obtain total cubic feet of grain by multiplying the length by the width by the average depth (all in feet). Divide by 1¼ (or multiply by 0.8) to find bushels. Round bin—multiply distance around the bin by the diameter by the depth of grain (all in feet) and divide by 5.

2. *To find the number of bushels of ear corn in crib:* Rectangular crib—multiply the length by the width by the average depth (all in feet) and divide by 2½ (or multiply by 0.4) to find bushels. Round crib—multiply the distance around the crib by the diameter by the depth of the corn (all in feet) and divide by 10.

3. *To find tons of loose hay in mow:* Multiply the length by the width by the height (all in feet) and divide by 400 to 525, depending on the kind of hay and how long it has been in the mow.

4. *To find tons of loose hay in stack:* Rectangular stacks—secure the over-throw, O (the distance from the ground, close to stack, on one side over the top of the stack to the ground on the other side); the width, W; and the length, L (all in feet). Additional rules for round stacks will be found on page 4 of U.S.D.A. Leaflet No. 72, published in February, 1931.

The contents in cubic feet may then be determined as follows:

(a) For low round-topped stacks
$$[(.52 \times O) - (.44 \times W)] \times W \times L$$
(b) For high round-topped stacks
$$[(.52 \times O) - (.46 \times W)] \times W \times L$$

Divide the number of cubic feet thus secured by the following cubic feet allowed per ton:

When settled:	*1 to 3 months*	*Over 3 months*
Alfalfa Hay	485	470
Timothy	640	625
Wild Hay	600	450

For clover hay use about the same as alfalfa or slightly higher (500 to 512).

5. *To find the number of tons of straw:* Follow the same method as is used with hay except that about twice as many cubic feet (900 to 1,000) are allowed per ton.

6. *Fodder and stover* are usually estimated on the acre basis, estimating the amount of corn in the fodder and allowing some additional value per acre for the stover.

Index